T0073646

A History of Mathematical Impossibility

A History of Mathematical Impossibility

JESPER LÜTZEN
Department of Mathematical Sciences, University of Copenhagen, Denmark

OXFORD
UNIVERSITY PRESS

Great Clarendon Street, Oxford, OX2 6DP,
United Kingdom

Oxford University Press is a department of the University of Oxford.
It furthers the University's objective of excellence in research, scholarship,
and education by publishing worldwide. Oxford is a registered trade mark of
Oxford University Press in the UK and in certain other countries

Impression: 1

Published in the United States of America by Oxford University Press
198 Madison Avenue, New York, NY 10016, United States of America

British Library Cataloguing in Publication Data
Data available

Library of Congress Control Number: 2022942881

ISBN 978–0–19–286739–1

DOI: 10.1093/oso/9780192867391.001.0001

Printed and bound by
CPI Group (UK) Ltd, Croydon, CR0 4YY

Preface

"Impossible" means many things. In everyday language it often means "very difficult" or "it is not yet known how to do it." In mathematics it means "it has been proved that it cannot be done." This book deals with the history of mathematical theorems stating that something is impossible in this strict sense of the word. Like many other mathematicians and laypersons, I have been fascinated by impossibility theorems for many years. One of the first courses I taught at Odense University, Denmark, dealt with the history of the classical problems: the quadrature of the circle, the duplication of the cube, and the trisection of the angle. Afterwards I published the lectures as a book (Lützen 1985), intended for Danish high schools. While working on the lectures and the book I was struck by the fact that the secondary sources dealing with the history of the classical problems had very little to say about the impossibility question. Many scholarly as well as popular books dealt with different constructions of the solutions to the problems in particular in ancient Greece; but when it came to the impossibility of constructing the solutions by ruler and compass alone, most books and papers (including my own) jumped from a few vague remarks about the Greeks to a treatment of the impossibility proofs form the nineteenth century. So when I returned to impossibility problems in Lützen (2008) I decided to start by looking into the impossibility of the classical problems. My research led to four papers (Lützen 2009, 2010, 2013, 2014) in which I tried to fill in the gaps in the history of the impossibility of these construction problems. Having finished these specialized investigations, I decided to write a book on the history of mathematical impossibility theorems in general. At that time I had become aware of the very interesting research by Davide Crippa that was partially published in his book (Crippa 2019) and of the rising interest in impossibility questions in science discussed in books like Barrow (1998) and du Sautoy (2016).

While I worked on my book, two related books appeared, namely Stillwell (2018) and Richeson (2019). I only became aware of the two books after I had sent my manuscript to Oxford University Press for publication. Some of the referees pointed my attention to them. Fortunately, my take on the problem of mathematical impossibility was rather different from Stillwell's and Richeson's.

Stillwell's *Yearning for the Impossible: The Surprising Truths of Mathematics* is a very captivating popular presentation of some of the great wonders of mathematics, such as the fourth dimension, the irrational, and the imaginary. I have focused more on the impossibility theorems. Richeson's delightful book *Tales of Impossibility* covers the three classical problems and the problem of constructing regular polygons. My book covers many other impossible problems. In addition, while Barrow's and Richeson's books are primarily popular books about mathematics using history as a means, my book focuses on the *historical* development of the problems. It is a book on the history of mathematics.

This book could serve as a textbook in a course on the history of mathematics as well as a leisurely introduction to the subject. Recently, the importance of impossibility questions in the mathematics classroom has been raised by Winicki-Landman (2007) and myself (Lützen 2013). The present book could enlighten the discussions in the classroom.

My aim has been to write a semi-popular book based on solid historical research. It should be readable by a mathematically interested audience with a good understanding of high school mathematics. In the text and in a separate list of recommended reading I have referred to many secondary sources where the reader can read more about the history of mathematics discussed in the book. I have also referred to many of the important primary sources discussed in the present book, in particular those sources that I quote. However, I have not included all the primary and secondary publications I have consulted because that would have made the list of references extremely long. By consulting the list of references in the papers and books mentioned in the bibliography of this book, the readers can easily work their way back to the primary sources. I have primarily referred to English translations if they exist.

Many colleagues have encouraged me while I have been working on this book. In particular, I wish to thank Davide Crippa, whose Ph.D. thesis I was so fortunate to examine and discuss with him. On that occasion, I also had the pleasure of discussing impossibility problems with Marco Panza. While I worked on the paper on Arrow's impossibility theorem I contacted Mark A. Satterthwaite. In an e-mail he explained how he had come to work on the so-called Gibbard–Satterthwaite theorem. I wish to thank him for this nice piece of information.

Many referees have given very helpful feedback on the first draft of the book. They have corrected mistakes and suggested additions that have improved the book. The following have revealed their names: Jeremy Gray, Reinhard Siegmund-Schultze, David Rowe, Jose Ferreiros, and Robin Wilson. I wish to thank them as well as the anonymous referees. Moreover, I wish to thank Jan

Hogendijk who helped me find material concerning impossibility statements in the medieval Arab world.

Illustrations are an important part of a book. In the process of searching for suitable pictures, I have received help from many corners of the world. First of all, Jim Høyer has made fine (and even improved) photographic reproductions of several of the portraits that we have in our archive at the Department of Mathematical Sciences at the University of Copenhagen. Pamela Moreland from the Stanford News Library has provided me with a picture of Kenneth Arrow, Cécile Bouet from the Bibliothèque de l'Institut de France has allowed me to use a picture of Joseph Liouville, Caitlin Rizzo from the Institute of Advanced Study in Princeton has provided me with photos of John von Neumann and Kurt Gödel, and Jennifer Hinneburg from the Mathematisches Forschungsinstitut Oberwolfach provided me with pictures of Georg Cantor, Eugenio Beltrami, Ferdinand Lindemann, and Carl Friedrich Gauss. I wish to thank all of the above for their help.

Finally I wish to thank Oxford University Press for their professional and kind transformation of my manuscript into a presentable book. In particular, it was a pleasure to work together with Dan Taber, John Smallman, Hayley Miller, Hariharan Siva, and Karthiga Ramu.

Jesper Lützen
Copenhagen
April 2022

Contents

1. Introduction 1
   - 1.1 The organization of the book 1
   - 1.2 What is an impossibility theorem? 4
   - 1.3 Meta statements and mathematical results 5
   - 1.4 Why are impossibility results often misunderstood among amateur mathematicians? 6
   - 1.5 Impossibility results in mathematics and elsewhere 9
   - 1.6 A classification of mathematical impossibility results 12
   - 1.7 Impossibility as a creative force 13
2. Prehistory: Recorded and Non-recorded Impossibilities 15
3. The First Impossibility Proof: Incommensurability 19
   - 3.1 The discovery 19
   - 3.2 The consequences of the impossibility theorem 23
   - 3.3 Incommensurable quantities in Euclid's *Elements* 25
4. Classical Problems of Antiquity: Constructions and Positive Theorems 27
   - 4.1 Squaring a circle 28
   - 4.2 Doubling the cube 35
   - 4.3 Trisecting the angle 39
5. The Classical Problems: The Impossibility Question in Antiquity 42
   - 5.1 Existence and constructability 42
   - 5.2 Pappus on the classification of geometric problems 44
   - 5.3 The quadrature of a circle 47
   - 5.4 Using non-constructible quantities: Archimedes and Ptolemy 48
6. Diorisms: Conclusions about the Greeks and Medieval Arabs 52
   - 6.1 Diorisms 52
   - 6.2 Conclusion on impossibilities in Greek mathematics 56
   - 6.3 Medieval Arabic contributions 59
7. Cube Duplication and Angle Trisection in the Seventeenth and Eighteenth Centuries 65
   - 7.1 The seventeenth century 65
   - 7.2 Descartes's analytic geometry 66
   - 7.3 Descartes on the duplication of a cube and the trisection of an angle 69

7.4 Descartes's contributions 74
7.5 The eighteenth century 75
7.6 Montucla and Condorcet compared with Descartes 79

8. Circle Quadrature in the Seventeenth Century 81
8.1 "Solutions" and positive results 81
8.2 Descartes on the quadrature of a circle 83
8.3 Wallis on the impossibility of an analytic quadrature of a circle 85
8.4 Different quadratures of a circle 89
8.5 Gregory on impossibility proofs and the new analysis 91
8.6 Gregory's argument for the impossibility of the algebraic indefinite circle quadrature 93
8.7 Huygens' and Wallis' critique of Gregory 96
8.8 Leibniz on the impossibility of the indefinite circle quadrature 99
8.9 Newton's argument for the impossibility of the algebraic indefinite oval quadrature 102
8.10 Why prove impossibility 105

9. Circle Quadrature in the Eighteenth Century 108
9.1 Joseph Saurin (1659–1737) 108
9.2 Anonymous 110
9.3 Thomas Fantet De Lagny (1660–1734) 110
9.4 The enlightened opinion 112
9.5 D'Alembert 114
9.6 The French Academy of Sciences. Condorcet 116
9.7 Enlightening the amateurs 117
9.8 Lambert and the irrationality of π 118

10. Impossible Equations Made Possible: The Complex Numbers 121
10.1 The extension of the number system: Wallis's account 121
10.2 Cardano's sophisticated and useless numbers 124
10.3 The unreasonable usefulness of the complex numbers 127
10.4 A digression about infinitesimals 130

11. Euler and the Bridges of Königsberg 133

12. The Insolvability of the Quintic by Radicals 140
12.1 Early results 140
12.2 Paolo Ruffini 145
12.3 Niels Henrik Abel 149

13. Constructions with Ruler and Compass: The Final Impossibility Proofs 155
13.1 Gauss on regular polygons 155
13.2 Wantzel 160
13.3 The quadrature of a circle 163

14. Impossible Integrals 168
14.1 Early considerations 168
14.2 Abel's mostly unpublished results 170
14.3 Joseph Liouville on integration in algebraic terms 171
14.4 Liouville on integration in finite terms 173
14.5 Liouville on solution of differential equations by quadrature 177
14.6 Later developments 178
14.7 Concluding remarks on the situation *c*.1830 179

15. Impossibility of Proving the Parallel Postulate 181
15.1 The axiomatic deductive method 182
15.2 The parallel postulate and the attempts to prove it 184
15.3 Indirect proofs: Implicit non-Euclidean geometry 187
15.4 Non-Euclidean geometry: The invention 190
15.5 The help from differential geometry of surfaces 194
15.6. Conclusions 197

16. Hilbert and Impossible Problems 199
16.1 Impossibility as a solution; rejection of *ignorabimus* 199
16.2 Hilbert's third problem: Equidecomposability 202
16.3 Hilbert's seventh problem 204
16.4 Hilbert's first problem 205

17. Hilbert and Gödel on Axiomatization and Incompleteness 210
17.1 The axiomatization of mathematics 210
17.2 Hilbert's second Paris problem 212
17.3 The foundational crisis 215
17.4 Gödel's incompleteness theorems 216
17.5 Hilbert's tenth Paris problem 221
17.6 Conclusion 223

18. Fermat's Last Theorem 225
18.1 Fermat's contribution 225
18.2 Nineteenth-century contributions 230
18.3 The twentieth-century proof 233

19. Impossibility in Physics 238
19.1 The impossibility of perpetual motion machines 238
19.2 Twentieth-century impossibilities in physics 242

20. Arrow's Impossibility Theorem 248
20.1 The theory of voting 248
20.2 Welfare economics 251
20.3 The Impossibility theorem 253
20.4 The Gibbard–Satterthwaite theorem 256

21. Conclusion 259
21.1 From unimportant non-results to remarkable "solutions" 259
21.2 From meta-statements to mathematical theorems 260
21.3 Different types of problems and different types of proofs 260
21.4 Pure and applied impossibility theorems 262
21.5 Controversies 263
21.6 Impossibility as a creative force 263

Recommended Supplementary Reading 265
References 267
Index 279

1
Introduction

Theorems stating that something is impossible are among the most famous theorems in mathematics. The proof of Fermat's last theorem is the latest widely celebrated proof of such a theorem. Earlier examples include the impossibility of solving the quadrature of a circle by ruler and compass, the impossibility of proving the parallel postulate, the impossibility of solving the general quintic equation by radicals, and Gödel's proof that it is impossible to give an axiomatic basis for, e.g., arithmetic such that all statements in the theory can be proved or disproved.

This book deals with the historical development of such impossibility results. I shall show that the importance ascribed to such theorems has changed over time. I shall argue that there is a hierarchy of different kinds of impossibility results and that each kind has first been treated as a meta-mathematical problem before it entered into mathematics proper. Moreover, I shall analyze the differing roles played by impossibility results in the architecture of mathematics and emphasize the creative role of impossibility theorems in extending mathematical discourse and theory.

1.1 The organization of the book

The book is organized as follows: In the introduction, I shall address some general questions about impossibility theorems such as the following: Can one meaningfully speak of a class of impossibility theorems as opposed to "positive" theorems? Are impossibility theorems unique to mathematics? Why are many of these results puzzling and hard to understand or even accept for non-mathematicians? I shall introduce a hierarchy of impossibility theorems and discuss the distinction between mathematics proper and meta-mathematics, both of which will play an important role in the subsequent historical narrative. Finally, I shall call attention to a strategy widely used by mathematicians to circumvent impossibilities by altering the rules of the game.

The subsequent chapters will examine the history of a selection of prominent impossibility theorems in more or less chronological order. We shall

A History of Mathematical Impossibility. Jesper Lützen, Oxford University Press.
 DOI: 10.1093/oso/9780192867391.003.0001

begin with the traces in pre-Greek texts of an awareness of impossibilities in mathematics. Then we shall turn to the first recorded proof of impossibility, namely the proof of incommensurability of certain line segments attributed to the Pythagoreans. Remaining in ancient Greece we shall then deal with the so-called classical problems, i.e., quadrature of a circle, duplication of a cube, and trisection of an angle. The Greeks found various constructions using more or less complicated curves or instruments but continued to ask themselves whether simpler means were possible. In particular, constructions by ruler and compass were sought for. However, gradually the majority of mathematicians became convinced that at least the duplication of a cube and the trisection of an angle, and possibly also the quadrature of a circle, were impossible with these restricted means. However, we find little trace among the ancients of any desire to prove the impossibility mathematically. The desirability of such impossibility proofs only dawned on mathematicians in the seventeenth century.

Before we turn to these impossibility investigations we shall discuss another Greek impossibility-related topic, namely diorisms. They are conditions for the solvability of problems that are solvable in certain cases and unsolvable in other cases. In this connection we shall briefly consider how the medieval Arab mathematicians continued the discussion of impossible problems and formulated new and interesting diorisms.

The inconclusive seventeenth and eighteenth century discussions concerning the (im)possibility of the classical problems is the intriguing subject of Chapters 7–9. Some of the foremost mathematicians debated whether the quadrature of a circle could be done analytically, but they did not agree on the exact meaning of this question nor its relation to the problem of construction by ruler and compass. The use of analytic methods led to the ultimately misguided belief that the impossibility of the duplication of a cube and the trisection of an angle had been provided with rigorous proofs.

The definite proofs of the impossibility of constructing the classical problems with ruler and compass will be discussed in Chapter 13. The nineteenth-century proofs relied on the theory of equations developed immediately before. So before the grand finale of the impossibility of the classical problems we shall discuss two impossibility questions concerned with the problem of solving equations (Chapters 10 and 12). The first one concerns the fundamental theorem of algebra (Chapter 10). Here the impossibility statement (that certain polynomial equations have no real solutions) is trivial. The interesting thing is to see how mathematicians circumvented the impossibility by inventing the complex numbers. This may be the most important instance of the use of imaginary or ideal elements to overcome a mathematical impossibility.

Before we turn to the second problem concerning equations we shall in Chapter 11 discuss Euler's treatment of the problem of the bridges of Königsberg. His proof of the impossibility of a so-called Euler tour over the bridges is usually considered the early beginning of graph theory.

The second problem about equations concerns their solvability by radicals (i.e., n'th roots) (Chapter 12). The eighteenth-century research in this area led to a great many new results in algebra, culminating with the proof of the impossibility of solving the general quintic (fifth degree) equation by radicals. The new algebraic methods also turned out to be powerful tools in other proofs of impossibility. The classical problems (Chapter 13) and the impossibility of expressing certain integrals in finite form (Chapter 14) were among the most spectacular.

In Chapter 15 we turn to the parallel postulate. Already in antiquity, mathematicians tried to get rid of it as a postulate by proving it from the other postulates. If they had succeeded the postulate would have turned into a theorem. However, in the nineteenth century it was shown that it is impossible to find such a proof. On the way to this impossibility theorem, non-Euclidean geometries, as well as a general method for proving the impossibility of deducing a particular statement from a given set of axioms, were invented. Both inventions revolutionized mathematics (and physics).

Around 1900, Hilbert was one of the first mathematicians to emphasize the importance of impossibility results (Chapter 16). He was convinced that all mathematical problems could either be solved or be proved insolvable within a mathematical theory determined by a suitable system of axioms. However, Gödel in 1930 proved him wrong (Chapter 17), establishing that no axiom system is strong enough to prove all theorems in number theory. Thus while emphasizing impossibility results, Hilbert overlooked the next spectacular impossibility theorem in mathematics. Gödel's results had important consequences for the foundation and philosophy of mathematics.

No book about mathematical impossibility theorems can avoid an account of the latest spectacular impossibility result, Fermat's last theorem formulated by Fermat in the seventeenth century and proved only at the end of the twentieth century. A brief and non-technical account of its history is contained in Chapter 18. It is a story of a famous problem that may not in itself be of the greatest importance, but its solution inspired the development of many important mathematical theories and methods.

The above-mentioned impossibility results all concern very pure parts of mathematics. We shall end with impossibility results in applied mathematics exemplified by physics and social science. In physics we shall consider two examples: First, the impossibility of constructing a perpetual motion machine and its relation to the first and second laws of thermodynamics; Second is von Neumann's proof that one cannot avoid the fundamental statistical nature of quantum mechanics by introducing hidden variables. In the social sciences we shall consider Arrow's impossibility theorem stating, loosely speaking, that it is impossible to design a just election method in elections with more than two candidates (Chapter 20).

In the Conclusion, I shall briefly recapitulate what we have learned about the history of mathematical impossibility results over the past 2000 years. What were their status at various periods, what did they contribute to the development of mathematics?

1.2 What is an impossibility theorem?

From a logical point of view, it makes no sense to single out a particular class of mathematical impossibility theorems. Any theorem can be formulated in either a positive or a negative way.[1] Take, for example, Fermat's last theorem: It is impossible to solve the equation $x^n + y^n = z^n$ when we assume that x, y, z, and n are integers and $n > 2$. Formulated in this way, it is an impossibility theorem. But it can also be formulated in a positive way: $x^n + y^n \neq z^n$ for all integers x, y, z, and $n > 2$. Similarly, Pythagoras's theorem stating that in a right-angled triangle the square on the hypotenuse equals the sum of the squares of the two other sides could also be formulated in the negative as an impossibility theorem: It is impossible to construct a right-angled triangle on three line segments unless the square of the longest line segment equals the sum of the squares on the two shorter line segments. In general any positive theorem can be formulated as an impossibility theorem by saying that the negation of the theorem is impossible.

Yet, in the development of mathematics, some theorems have been formulated as impossibility theorems while others have been formulated as positive statements. Fermat's last theorem is almost always formulated as an impossibility theorem, whereas Pythagoras's theorem is always formulated in a positive way. Psychologically the two types of theorem simply feel different. Why is that?

[1] Many of the points of this Introduction were published in Danish in Lützen (2008).

In an attempt to answer this question, note that mathematics, in addition to being a theorem-proving activity, also deals with problem-solving. In the domain of problem solving there is a great difference between finding a solution and finding a proof that the problem cannot be solved. An impossibility theorem is usually a theorem saying that a certain problem cannot be solved and this is why it can be distinguished from a positive result. Important impossibility theorems are those that show the unsolvability of a problem that for some reason or another one would assume to be solvable.[2]

Impossibility theorems as well as positively phrased theorems come in many degrees of difficulty. Some are trivial while others are very difficult and surprising. In a deductively organized presentation of mathematics, one begins by proving rather simple consequences of the axioms and only gradually reaches interesting results. In principle, this holds for both impossibility theorems and positive statements. For example, in order to show that the side and diagonal in a square have no common measure (are incommensurable) one usually gives an indirect proof that is based on the theorem that it is impossible that an integer be both even and odd. However, this last statement is so trivial that no mathematician would bother to formulate it as a theorem. Indeed, trivial impossibility results rarely make it into mathematics texts. The impossibility results discussed in this book are the hard ones, those that have become famous either because they are counterintuitive or because they have been very difficult to prove.

1.3 Meta statements and mathematical results

Mathematicians have always been engaged in two types of discourse. One is the mathematical discourse in which mathematical problems are solved and (since the ancient Greeks) theorems are formulated. This discourse is governed by strict rules. In particular, it is required that theorems are proved. But mathematicians also engage in another discourse *about* their science. This discourse is often called meta-mathematics. It deals with methodological questions such as the relative importance of mathematical theories or theorems, the preferable methods of proof, the preferable direction of research, and philosophical, sociological, and historical questions concerning mathematics. In this discourse, one may give arguments but rigorous proofs are not a required part of the game.

[2] In this book I shall use the phrase "impossible problem" to designate a problem that cannot be solved in a specified way.

One of the aims of this book is to convince the reader that many impossibility results emerged as meta-statements and only subsequently developed into true mathematical theorems. This argument is put forward in order to explain why many impossibility theorems were not at first considered to be in need of proofs. Indeed, if an impossibility statement was considered as a meta-statement it would not call out for a proof any more than other meta-statements such as "this is an important theorem" or "this is the most beautiful proof I have ever seen."

In Section 1.4, I shall argue that it is quite natural to interpret impossibility statements as meta-statements *about* mathematics rather than as statements *within* mathematics. Once we have realized this, we also see that rather than posing the Whig question—Why did earlier mathematicians not provide proofs of their impossibility statements?—we should marvel: How did it come about that impossibility results about mathematics were transformed into theorems in mathematics? More generally: What allowed mathematics to develop into a science that can deal with its own methodology and in particular can prove theorems about its own limits? I do not claim that this book answers this large question completely, but I shall argue that it is a question worth asking and a question that is relevant to the history of impossibility theorems.

1.4 Why are impossibility results often misunderstood among amateur mathematicians?

Impossibility theorems have traditionally been a favorite subject among amateur mathematicians. In particular, the three classical construction problems, the quadrature of a circle, the duplication of a cube, and the trisection of an angle, have attracted the interest of many intelligent but untrained persons seeking immortal fame. Why do so many people waste their lives trying to refute these impossibility theorems? In many cases it is because they have not understood what is claimed to be impossible. In the case of the classical construction problems the theorem states that exact constructions are impossible with ruler and compass in a finite number of steps. If these limitations are dropped, the impossibility statements seem obviously false and cry out for refutation.

However, even people who have understood the formulation of the impossibility theorems often try to find constructions. Though I do not have any statistics to back it, I am sure that there are many more amateurs trying to find impossible constructions than there are amateurs who try to disprove other theorems in mathematics. Why? One answer could be that the proofs of the impossibility theorems are so complicated that most amateurs cannot follow them. However, there are many appealing positive theorems with equally difficult proofs that are nevertheless accepted by amateurs. I will suggest that impossibility theorems are so attractive targets for amateurs because they misunderstand the nature of the impossibility claim.

In ordinary language "impossible" means something like "very difficult," "no one has been able to do it," or something similar. There are plenty of stories about people who have been able to do what has been declared impossible. In fact, in a database of quotes posted on the internet (Zaadz 2007) most of the quotes involving the word "impossible" have the following clear message: Nothing is really impossible. Do not believe the experts when they declare something to be impossible. If one wants to achieve something in life one should go for the things that have been declared impossible. Here are some typical quotes:

> "This has always been a motto of mine: Attempt the impossible in order to improve your work."
>
> (Bette Davis)
>
> "It is difficult to say what is impossible, for the dream of yesterday is the hope of today and the reality of tomorrow."
>
> (Robert H. Goddard)
>
> "It's kind of fun to do the impossible."
>
> (Walt Disney)
>
> "Nothing is impossible to a willing heart."
>
> (John Heywood)
>
> "Nothing is impossible: There are ways which lead to everything: and if we had sufficient will we should always have sufficient means. It is often merely for an excuse that we say that things are impossible."
>
> (Duc de La Rochefoucauld)
>
> "Most of the things worth doing in the world had been declared impossible before they were done."
>
> (Louis Dembitz Brandeis)
>
> "The young do not know enough to be prudent, and therefore they attempt the impossible, and achieve it, generation after generation."
>
> (Pearl S. Buck)
>
> "Every noble work is at first impossible."
>
> (Thomas Carlyle)

"Things are only impossible until they're not."

(Jean-Luc Picard)

"There is nothing impossible to him who will try."

(Alexander the Great)

"Man is the only creature that strives to surpass himself, and yearns for the impossible."

(Eric Hoffer)

"The impossible is often the untried."

(Jim Goodwin)

"Some of the world's greatest feats were accomplished by people not smart enough to know they were impossible."

(Doug Larson)

"I love those who yearn for the impossible."

(Johann Wolfgang von Goethe).[3]

With the encouragement of such a selection of dignitaries, it is not surprising that many will try to do the impossible also in mathematics. However, what amateur mathematicians seem to overlook is that in modern mathematics, impossibility claims play a different role than they do in other areas of life. Outside of mathematics a problem and the claim of its impossibility usually have very different natures. Consider, for example, the following quote from the same list as those above:

"The Wright brothers flew right through the smoke screen of impossibility."

(Charles Kettering)

To construct an airplane is an engineering problem, whereas a claim of the impossibility of such an endeavor, though it is about engineering, cannot be established through a technical construction. It is so to speak a meta-engineering statement. Its claim to truth is of a different nature than an actual construction of an airplane.

In modern mathematics, on the other hand, an impossibility theorem has the exact same nature as any other theorem. In particular, it is proved according to the same methods as any other mathematical theorem. Amateur mathematicians seem to believe that impossibility theorems in mathematics are meta-theorems just as many other impossibility claims and that they have weaker claims to truth than ordinary mathematical theorems.

[3] The point here is not whether the quotes are accurate and attributed to the right people, but what this selection says about the popular interpretation of the meaning of the impossible.

This would explain why amateur mathematicians are more prone to disproving impossibility theorems than other mathematical theorems. They know the force of proofs in mathematics and so even if they have not actually followed a proof of Pythagoras's theorem they know that when a proof exists there is no point in trying to disprove the theorem. However, misled by the role of impossibility statements in other walks of life, they may very well mistake an impossibility theorem for a meta-mathematical statement which is about mathematics, but is not itself a mathematical theorem. If the impossibility claim is only a meta-mathematical statement it may have the same status as other impossibility statements and these are, as we have seen in the quotes, claims that await refutation. One often hears the argument: How do mathematicians know that no construction, however complicated, can lead to the solution of the classical problems? Surely, they cannot have tried them all! On the other hand, one does not hear the following question: How do mathematicians know that Pythagoras' theorem is true for all right-angled triangles? Surely, they have not investigated them all! The reason could be that amateur mathematicians know that Pythagoras's theorem is a theorem and thus has a proof, whereas they mistake the impossibility theorem for a meta-theorem that does not have a proof.

If I am right, it is no wonder that many circle squarers think that mathematicians are arrogant when they claim that a circle quadrature is mistaken even before they have cast a glance on the construction.

1.5 Impossibility results in mathematics and elsewhere

In Section 1.4, we contrasted the mathematical and the popular understanding of the word impossible. Before we enter into the historical discussion, it may be worth contemplating if well-established impossibility results are unique to mathematics or whether other sciences or other areas of life have such results. Indeed, the question of impossibility in areas outside of mathematics has been taken up recently by several scientists-philosophers. In a book entitled *No Way: The Nature of the Impossible* (Davis and Park 1987) a series of experts point to impossibilities in fields ranging from mountain climbing and medicine to physics, computer science, mathematics, and philosophy. More recently the fundamental question of what we can know about the world and what is impossible to know has been discussed in the books *Impossibility: The limits of Science and the Science of Limits* by cosmologist John D. Barrow (1998) and *What We Cannot Know: From Consciousness to the Cosmos, the Cutting Edge of Science Explained* by mathematician du Sautoy (2016). Similarly, the computer scientist Joseph F. Traub has argued: "My goal is to move the distinction between the unknown and the unknowable from philosophy to science and thereby enrich science" (Traub 2007).

Does that mean that we are witnessing the beginning of a development where impossibility statements in science begin to play a role similar to those of mathematics? In one sense, the answer may be yes: Both Traub and Barrow argue that it is important for science to deal with its own limits, and that it is important to do so with scientific means. This development is similar to the development in mathematics in which meta-claims about impossibility in mathematics have been turned into proper mathematical impossibility theorems that are proved with mathematical methods. In another way, however, the answer seems to be no. Indeed as pointed out by Traub: "We cannot prove scientific unknowability. That can only be done in mathematics" (Traub 2007).

Let us briefly analyze the different types of impossibility claims that one can find in science: As far as I can see, there are essentially three types of impossibility results in science. The first type are the purely empirical impossibility results. The second type relies on a mathematical description of the world and the third deals with more fundamental philosophical impossibilities mostly related to the range of scientific inquiry.

The empirical impossibilities range from statements like "No swan is black" to "it is not possible to build a perpetuum mobile." The first of the impossibility statements is (or was, when it was a feasible statement) of the same type as most non-scientific impossibility statements and an exploring naturalist would be wise to try and disprove it.

The nonexistence of a perpetuum mobile is also a claim based on empirical evidence but still has a different status. The reason is that it is a (central) part of a fundamental mathematical model of the physical world. In this way, it is also an example of the second type of impossibility theorems in science. Other famous impossibility theorems of the second type include statements like "energy can never be created," "no signal can travel faster than light," and "one cannot determine both the place and the momentum of a particle precisely" (see Chapter 19). Such impossibility statements can be proved with mathematical methods within the given model of nature. This gives them a better foundation than more contingent impossibilities like the impossibility of black swans. However, the difference is one of degree rather than one of principle. Indeed the mathematical proofs only establish the impossibility theorems as mathematical impossibility theorems. Their relation to nature is no more solid than the relation between nature and the mathematical model in which they are formulated. Thus, in so far as the mathematical model is a result of empirical evidence, this type of impossibility results are also of an empirical nature. Also social science have impossibility results of this type, for example Arrow's impossibility theorem (see Chapter 20).

The third type of impossibility statements is the philosophical ones like "time travel is impossible" or "there cannot exist an omnipotent being." This type of theorems has proofs just as the mathematical ones. And like the mathematical impossibility theorems the proofs are often indirect: If I could travel back in time I could kill my own mother before she gave birth to me. This is impossible QED. If there existed an omnipotent being (s)he could create a stone so big that (s)he could not lift it, and then (s)he would not be omnipotent after all. A particularly important impossibility result is the skeptic argument stating that it is impossible to obtain (certain) knowledge inductively. This would imply that impossibility results of the first two types are impossible.

As far as I can see such philosophical theorems and their proofs are either too vague to be real theorems or they can be made precise, in which case they will be of the same kind as mathematical impossible theorems. They will be the result of a logic analysis and so only depend on the accepted rules of inference and the precise axiomatic structure within which the premises and the conclusion are formulated.

Thus, as far as I can see there are two fundamentally different types of impossibility theorems: The empirical ones and the mathematical-logical ones. The difference between them is the same as the general difference between empirical sciences and mathematical sciences. Thus in this respect there is no reason to single out impossibility theorems as a special type of theorems. Both mathematical-logical and empirical impossibility theorems are characterized by the same type of strengths and weaknesses as any other theorem of these types. In particular, though the mathematical impossibility theorems are free of the problems surrounding empirical inductive sciences, and therefore can be considered more absolute than other impossibility statements, they are not immune to the usual discussion about what a mathematical proof proves. Nor are they immune to human errors.

The above argument rests on a modern and logical conception of mathematics in which there is no logical difference between impossibility results and positive results. Still, even modern mathematicians have a rather clear feel for the difference between the two types of results, and as we shall see, the difference was considered much greater in former times.

1.6 A classification of mathematical impossibility results

In order to understand which types of mathematical impossibility problems and theorems were formulated and solved at various points in history, it is helpful to classify them.

First, I shall classify them according to what it is that is claimed and proved to be impossible. The conceptually simplest type is the theorems stating that a certain kind of object such as a number or a geometric figure does not exist. Examples include Fermat's last theorem. The next type states that a certain type of process does not exist. A good example of such an impossibility theorem is the insolvability of the quintic[4] by radicals. Here the fundamental theorem of algebra guarantees the existence of five roots of the quintic, but the theorem states that these solutions cannot in general be found through a procedure involving only the rational operations and root extractions. On the next level of conceptual sophistication come the theorems stating that certain theorems cannot be proved. The impossibility of proving the parallel postulate is the first such theorem. Finally, the conceptually most complex type of impossibility results are those stating that certain properties of axiomatic structures cannot be proved. Gödel's incompleteness theorem is a case in point.

This classification goes from the simple object level toward more and more reflective meta levels. If we accept that mathematics historically started as a method to deal with real objects numerically and geometrically and only gradually became a methodologically reflexive science, it is not surprising that the first impossibility results that were formulated and proved belong to the first category, the next to the second category, etc. Among the first two of the above-mentioned types of impossibility results, one can also distinguish between the finite ones and the infinite ones. The finite ones state the nonexistence of a special type of object or process among a finite number of possibilities. If this number is small and easy to survey, these impossibility results are trivial and as a rule not even formulated in the literature. For example that 11 is a prime (i.e., has no divisors except 1 and 11) is seldom mentioned as a theorem. Only theorems stating the impossibility among an infinity of objects or processes (or a large number complicated to survey) are interesting.

Finally, there seems to be a difference between the instances where one can get as close as one wishes to the impossible thing and the instances where one cannot even get close to it. When one can get as close as one wishes, as in the case of the classical problems or the solution of equations by radicals, one may

[4] The quintic is the polynomial equation of degree 5.

mistake an approximate solution for an exact one, or the approximations may suggest that an exact solution is within reach. More subtly, one may explicitly or implicitly include limiting procedures among the allowed procedures in the solution of the problem (we shall see explicit examples of this in Chapter 8). In number theory, on the other hand, an equality is either exactly satisfied or it is off by at least one, so here it is intuitively easier to accept that an identity can never be satisfied.

1.7 Impossibility as a creative force

At first sight, an impossibility may appear as a roadblock on the road to further progress of mathematics. However, historically most impossibility results have been incentives to progress. The reason is that mathematicians behave like ordinary people who challenge claims of impossibility. They get around the impossibility by changing the rules of the game. In everyday claims of impossibility, the rules are usually not formulated so clearly that the change is noticed. For example in the case of the Wright brothers the impossibility they flew through was a statement that a person cannot fly. Clearly, the original meaning of the statement was that a person cannot get off the ground by flapping his or her arms. The introduction of an airplane changes the rules of the game, but since the rules were not formulated explicitly in the first place, it looks as though the pilots have disproved an impossibility. When Alexander the Great cut the Gordian knot, he used a similar strategy.

Mathematicians try to be more precise in their formulations of impossibility claims. For example, when it is said that it is impossible to square the circle, it is stated explicitly that only ruler and compass can be used. It is even specified exactly what one is allowed to do with these two instruments. Therefore, in principle, mathematicians who want to circumvent the roadblock of an impossibility theorem must explicitly change the rules. In practice, however, even mathematicians sometimes use implicit assumptions or rules that can be bent. For example, prior to 1550 it would seem uncontroversial to claim that the quadratic equation $x^2 + 1 = 0$ has no solutions. The proof would even seem trivial. Only after the complex numbers were invented did one realize that the intended correct formulation of the impossibility statement should have been "There is no *real* solution to the said equation."

Whether the rule-changes induced by impossibility statements were explicit or implicit, they have often been important in the progress of mathematics. One of the most used strategies around an impossibility result is to extend the field of discourse. The introduction of the complex numbers is an important example, as is the acceptance of other instruments or curves in the construction of the classical problems. However, there are also examples of different ways to get around impossibilities. The most striking example is the Greek way around the problem of incommensurability. They avoided the problem by decoupling geometry from arithmetic, creating an entirely new pure form of geometry.

We should also remark that in many cases the way around an impossibility result was already in place before the impossibility was formulated and proved in the first place. For example, many different solutions to the classical problems were known long before the Greeks began to doubt their possibility with ruler and compass, and even longer before proofs of impossibility were put forward in the nineteenth century. Similarly, numerical or geometric methods of solving equations were around long before the wish for solutions by radicals was formulated and even longer before the impossibility results in this area were proven. In such cases, the impossibility came about as a result of a hierarchy of the known solution methods, some of which seem better than others in some way (simpler, more accurate, methodologically preferable, ...). We will see that such hierarchies were formulated in late antiquity and in the seventeenth century. They automatically lead to impossibility questions. Indeed, given a hierarchy of more and more general (but less and less desirable) solution methods $S_1, S_2, \ldots, S_n, \ldots$, it is natural to divide problems into classes: a problem belongs to class m if it can be solved by the methods S_m but cannot be solved using methods S_{m-1}. The latter of these requirements is an impossibility statement.

2
Prehistory
Recorded and Non-recorded Impossibilities

Problems of a mathematical nature may be impossible to solve with given means. This is an insight that was certainly acquired by humans long before they could record it for later generations. For example, they must have discovered that it is impossible to divide a herd of 7 sheep evenly between 3 persons, at least if the sheep should remain unharmed. The similar problem of dividing 7 loaves of bread evenly among 3 people, on the other hand, can be solved if one allows one of the loaves to be cut in three. This exemplifies a general principle in the development of mathematics and other areas of life: If a problem cannot be solved with some means, one tries to use more forceful means. In this case the trick is to allow the use of a knife. In mathematical terms one admits fractions in addition to the natural numbers (1, 2, 3, ...).

For obvious reasons our knowledge of prehistoric mathematics is somewhat speculative but already the very first written mathematical sources from Mesopotamia (Babylon) and Egypt (around 2000 BC) bear witness to the awareness of the impossibility question. These mathematical texts are all collections of problems with solutions. In most cases the problems have been carefully constructed so that they allow a solution of a particular simple kind, e.g., in terms of natural numbers. It is clear that the scribes of the texts were aware that many problems were not solvable with certain means. For example, an Egyptian problem asks the reader to determine a heap (think of an x) such that the heap and a fourth of the heap together makes 15.[1] The scribe correctly explains how to find the solution ($x =$) 12 but it is quite clear that he knew that if he chose numbers other than one-fourth or 15 he might need fractions to express the solution. The Egyptians introduced fractions written as sums of different unit fractions (fractions of the form $1/n$, where n is a natural number) and seems to have realized that first-degree problems of the above kind can always be solved using these means. They probably had no real argument

[1] This is problem 26 of the longest preserved mathematical text from ancient Egypt, the so-called Papyrus Rhind from about 1800 BC (Chase 1967). See Imhausen (2016) for more information on Egyptian mathematics.

A History of Mathematical Impossibility. Jesper Lützen, Oxford University Press.
 DOI: 10.1093/oso/9780192867391.003.0002

for this and it is unclear whether they knew that other problems, such as the determination of the area of a circle, could not be solved exactly using only fractions.

In Mesopotamia where mathematics was written in cuneiform symbols on clay tablets, fractions were expressed in another manner similar to our decimal fractions. The only difference is that they used a number system with base 60 instead of our base 10. When we write 2.583 we mean $2+5/10+8/100+3/1000$. Similarly, a Babylonian scribe who wrote 2;43,27,13 (using, of course, other symbols for the natural numbers) meant $2+43/60+27/60^2+13/60^3$. These so-called sexagesimal fractions provided a very efficient way of handling fractions. Still, the Babylonians discovered that there are, in fact, simple problems of division that are impossible using sexagesimal fractions. Indeed, in many tables of reciprocals, i.e., tables of the values of $1/n$ for n = 1, 2, 3, 4, ..., we find curious gaps. The tables begin with the values ½ = 0;30, ⅓ = 0;20, ¼ = 0;15, ⅕ = 0;12, $^1/_6$ = 0;10, but then most tables skip $^1/_7$ and continue with $^1/_8$ = 0;7,30. Of the fractions $1/n$ for n between 2 and 20 most tables skip the fractions $^1/_7$, $^1/_{11}$, $^1/_{13}$, $^1/_{14}$, $^1/_{17}$, $^1/_{19}$. The special property of these fractions is that they cannot be expressed in a finite sexagesimal fraction. They behave in the sexagesimal system just as ⅓ = 0.333333 ... behaves in the decimal system. Their sexagesimal fraction never ends. It is easy to show that $1/n$ has an infinite sexagesimal fraction if and only if n is divisible by a prime number other than 2, 3, and 5 (the prime divisors of 60). The Babylonian scribes may have noticed that, but they probably had no proof of it.[2]

The Babylonians used their tables of reciprocals to perform division. Indeed, in order to calculate a/b they looked up $1/b$ in a reciprocal table and multiplied the result by a, using a multiplication table. However, this method is impossible to use if $1/b$ is one of the fractions left out from the reciprocal table.[3] For example, in a text from about 1800 BC,[4] the scribe wants to divide 5;30 by 11. The text goes as follows: "The reciprocal of 11 *does not work*. What shall I take together with 11 to get 5;30? 0;30 ... ."[5] So the scribe tried to find 1/11 in the table of reciprocals and found that it is absent. So instead he guessed the

[2] Indeed, the preserved Babylonian tablets reveal no awareness of the concept of a prime number. This concept was introduced by the Greeks.

[3] Some tables of reciprocals contained approximations to the fractions that could not be expressed by a finite sexagesimal fraction.

[4] This is one of the oldest surviving Babylonian cuneiform clay tablets. It has the number BM 13901, indicating that it is kept at the British Museum. See Neugebauer (1951) and Høyrup (2002) for more information of Babylonian mathematics.

[5] The tablet is damaged in the place where the reconstruction has "does not work." However, from the rest of the text there seems little doubt about the reconstruction.

number that multiplied with 11 gives 5;30 (or 5½). It is easy to guess that the desired number is ½, i.e., 0;30.

This Babylonian text may be the first recorded example of an explicit impossibility claim in the history of mathematics. Together with the reciprocal tables it seems to suggest that the Babylonians were aware of the fact that 1/11 cannot be determined exactly as a finite sexagesimal fraction.

The above-mentioned calculation is in fact the end of a solution of a quadratic equation. It is remarkable that already 4000 years ago the Babylonians solved quadratic equations using a method that corresponds completely to our modern formula for the solution. In particular, the method asked for the extraction of the square root of the discriminant. When trying to formulate such quadratic equations the Babylonian scribes must have discovered that they needed to choose the given numbers (the coefficients) carefully if they wanted this square root be a natural number or a finite sexagesimal number as it happens in almost all the known Babylonian quadratic problems. The Babylonian scribes must also have discovered that some problems of this type were entirely impossible. For example, a standard problem on Babylonian tablets asks for the determination of the sides of a rectangle when the area and the half-circumference of the rectangle are known. This problem is only possible if the given area is smaller than or equal to the square on ¼ of the circumference. Otherwise, the discriminant is negative. It is very likely that the Babylonian scribes discovered these distinctions between possible and impossible problems.

Still, the scribes who faced these questions whenever they formulated problems for their pupils left no trace of their considerations about possible and impossible problems. To be sure, Olaf Schmidt (1980) has argued that the aim of the famous list of Pythagorean triples[6] in the tablet Plimpton 322[7] was to facilitate the formulation of solvable quadratic equations. Even if he is right, the scribes did not explicitly formulate this aim on the surviving tablets. This may be a result of the incomplete transmission of the sources, but it may also be a sign of a choice made by the scribes about what to include on the tablets. The surviving tablets predominantly contain practical (or seemingly practical) problems and their solutions. Perhaps the scribes thought that methods for formulating solvable problems were not of this kind and did not belong on the tablets. They may have been considered as meta-scribal rules, tricks of the trade that were best left to oral communication.

[6] That is, integer solutions a, b, c to the equation $a^2 + b^2 = c^2$.
[7] See Neugebauer and Sachs (1945, 38–41).

There are examples of Babylonian problems that cannot be solved exactly in sexagesimal fractions. For example, on a small tablet[8] dealing with the side and the diagonal of a square one finds the approximation 1; 24, 51, 10 of $\sqrt{2}$. All the three given sexagesimals are correct, but the Babylonians surely knew that it was only an approximation. It is even possible that they guessed that it would be impossible to find the square root of 2 exactly as a sexagesimal fraction. A stronger version of this impossibility result was proved by the Greeks.

[8] This clay tablet (YBC 7289) is preserved at the Yale Babylonian Collection. It was published in Neugebauer and Sachs (1945, 43).

3

The First Impossibility Proof

Incommensurability

3.1 The discovery

The ancient Greeks revolutionized mathematics by insisting that mathematical results must be demonstrated by a logically rigorous proof. It is debated whether the first Greek mathematician we know of, Thales (*c*.600 BC), gave proofs of the theorems attributed to him, but it seems certain that Pythagoras (*c*.500 BC) and his followers proved their mathematical results. Pythagoras initiated a philosophical brother- and sisterhood whose motto was "All is number," meaning that the essence of things is number (i.e., natural number) through which one can study them. The Pythagorean philosophy was the first attempt to understand the world in a mathematical way. It is said that Pythagoras came upon this idea through his experiments with musical instruments. In particular, he discovered that the harmonic musical intervals (the octave, the fifth and the fourth, etc.) were emitted when a string was divided in simple ratios 1:2, 2:3, or 3:4.

However, within a century it was discovered that even simple geometry could not always be described in terms of natural numbers. According to Pappus of Alexandria (*c.* AD 290–350), the first Pythagorean who divulged this imperfection of the Pythagorean philosophy was punished by the gods and drowned at sea. Other legends told by Iamblichus (*c.* AD 245–325) point to Hippasus of Metapontum (fifth century BC) as the unfortunate person. The problem he discovered is the so-called incommensurability of two line segments. Two line segments are called commensurable if there exists a small line segment (a measure stick) that measures each of them a whole number of times. If we call the two line segments a and b and the measure stick d, then we require that a is a whole number (say m) times d and b is another whole number (say n) times d. If this is the case, we call the small measure stick d a common measure of the line segments a and b. If two line segments

A History of Mathematical Impossibility. Jesper Lützen, Oxford University Press.
© Jesper Lützen (2022). DOI: 10.1093/oso/9780192867391.003.0003

have a common measure we can describe their ratio as a ratio between natural numbers as required by the Pythagorean philosophy:

$$\frac{a}{b} = \frac{md}{nd} = \frac{m}{n}.$$

What Hippasus or some other Pythagorean discovered was that there are pairs of line segments for which it is impossible to find a common measure. This impossibility shows that the ratio of the said line segments cannot be expressed as a ratio of natural numbers and this in turn was considered as an imperfection in the Pythagorean philosophy.

It is not clear from the existing sources which two line segments were shown to be incommensurable, nor is it mentioned which methods were used to prove it. One possibility is that it was the side and diagonal in a square that were proven to be incommensurable and that the method of proof was similar to the proof preserved in an interpolation at the end of Book X of Euclid's *Elements* (proposition 117). This proof was also mentioned by the philosopher Aristotle (384–322 BC) when he wanted to explain the method of proof by contradiction (Heath 1921, 91; Aristotle Anal. pr. i. 23, 41 a 26–7). The indirect method of proof that Aristotle connects closely with the first impossibility theorem is a method of argumentation used often also outside mathematics. If one wants to show that something cannot be, one shows that it has absurd consequences. In the case at hand, we assume that the side s and the diagonal d of a square are commensurable and show that it leads to a contradiction.

Theorem: *The side and diagonal in a square are incommensurable.*

Proof: Assume that the side s and the diagonal d are commensurable. As explained earlier, this would imply that their ratio could be expressed as a ratio between two natural numbers m and n:

$$\frac{s}{d} = \frac{m}{n}.$$

We can assume that the fraction m/n is reduced to lowest terms such that m and n have no common divisors. According to Pythagoras' theorem, that was in fact known to the Babylonians about one and a half millennia before Pythagoras, the square on the diagonal is twice the original square. In fact this also follows from a simple argument that Socrates makes a slave go through in Plato's dialogue *Meno* (Figure 3.1).

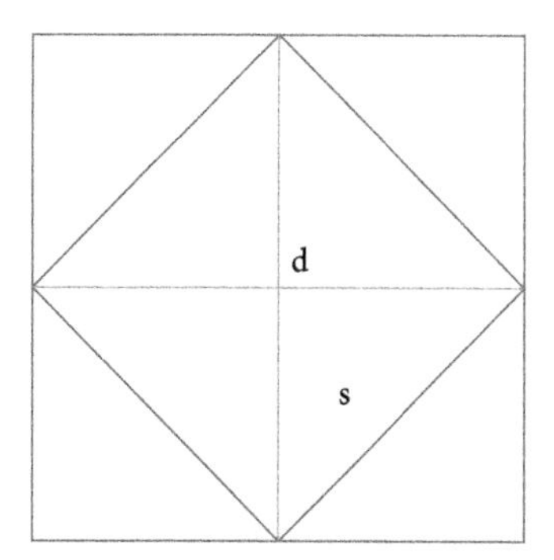

Figure 3.1 Illustrating Plato's argument that a square is half as big as the square on its diagonal

Thus,

$$\frac{m^2}{n^2} = \frac{s^2}{d^2} = \frac{s^2}{2s^2} = \frac{1}{2},$$

which implies that $n^2 = 2m^2$. From this, we conclude that n^2 is even and since the square of an odd number is odd we conclude that n itself must be even, that is, of the form $2p$ for some natural number p. Now since m/n was assumed to be reduced to lowest terms, *m must be an odd number.*

On the other hand if we insert $n = 2p$ into the equation $n^2 = 2m^2$ we get $n^2 = (2p)^2 = 4p^2 = 2m^2$ and thus $m^2 = 2p^2$. As above, this implies that *m is even* but we have also argued that it was odd. We have thus arrived at a contradiction and must conclude that our assumption that the side and the diagonal were commensurable must be false. Thus, they are incommensurable, as we wanted to demonstrate.

The use of the dichotomy even–odd makes this reconstruction a likely candidate for a Pythagorean proof. Indeed, as a symbol of the dichotomy female–male it played an important part in the somewhat mystical Pythagorean theory of numbers. However, another possible incommensurability proof has been put forward, namely a proof of incommensurability of the side and the diagonal in a regular pentagon using Euclid's algorithm. Speaking in favor of this proof as the original Pythagorean incommensurability proof is the legend that the pentagram was the holy symbol of the Pythagoreans. The pentagram is a five-pointed star consisting of the diagonals in a regular pentagon.

Euclid's algorithm, which is documented also before Euclid, is a method by which one can find the greatest common divisor of two numbers. You simply subtract the smallest number from the largest and you get a remainder. Then you consider the least of the two given numbers and the remainder and again subtract the smallest from the largest. Proceeding in this way one will eventually reach a remainder of zero and it is easy to prove that the last non-zero remainder will be the greatest common divisor of the two numbers. Now, if two line segments a and b are commensurable and if $a/b = m/n$ it is rather evident that Euclid's algorithm used on the pair of line segments a, b must run entirely parallel to Euclid's algorithm used on the pair of numbers m, n. That implies that it will end, and that the last non-zero remainder will be a line segment that measures both a and b a whole number of times. Therefore, if one can find two line segments whose Euclid algorithm does not end, one has found a pair of incommensurable line segments. And here the side a and the diagonal b in a regular pentagon might have presented themselves to a Pythagorean as simple examples (Figure 3.2).

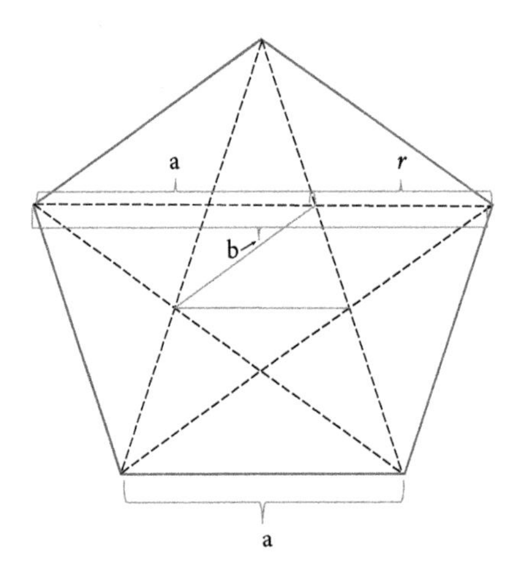

Figure 3.2 The side a and diagonal b in a regular pentagon and the remainder r. The dashed star shape is a pentagram

Indeed, if one subtracts the side a from the diagonal b one is left with the remainder r which is equal to the diagonal in the smaller regular pentagon formed by the diagonals. And if one subtracts this remainder from the side a of the original pentagon one is left with the side of the smaller pentagon. Thus, after two steps in Euclid's algorithm one has obtained the side and the diagonal of a smaller regular pentagon. After two more steps, one will be left with a side and a diagonal in an even smaller pentagon and so on. It is obvious that this process will continue ad infinitum and will never end with a common measure. Therefore, there is no common measure of the original side and diagonal: they are incommensurable.

Whether the original proof of incommensurability was one or the other of these two proofs or yet a third one is not so important. What matters is that the ancient Greeks discovered that there are pairs of line segments for which it is impossible to find a common measure. That means that it is not always possible to measure (in natural numbers) all line segments in a geometric figure. Thus, the Pythagorean doctrine "all is number" was untenable. This insight had a fundamental impact on the further development of Greek mathematics.

3.2 The consequences of the impossibility theorem

As pointed out in Chapter 2 it is possible that the Babylonians had already noticed (but not proved) that there are line segments whose lengths cannot be written exactly in their number system. As far as we can see from the sources, this did not bother them so much. They just determined approximations to the lengths. This was not an option for the Greeks who insisted that mathematical results be exact. In modern mathematics, we get around the problem of incommensurability by enlarging the domain of numbers. For example if the side of a square is 1 we denote the length of the diagonal by $\sqrt{2}$ and we show how one can calculate with such irrational numbers. The Greeks did not do that. They maintained that numbers were collections of units; i.e., they limited themselves to the natural numbers 1, 2, 3, ...[1] Therefore they distinguished sharply between continuous quantities such as line segments, two-dimensional figures, angles, etc., and discrete quantities like natural numbers.

In all other early mathematical cultures (Mesopotamia, Egypt, China, and India) geometry mostly dealt with the determination of the magnitude of geometric quantities measured as a number of units. Thus, calculations with numbers were an important part of their geometry. The discovery of incommensurable geometric quantities made the Greeks aware that such an approach to geometry was inexact. Instead, they developed a new style of geometry in which geometric quantities were not measured by numbers but were considered as magnitudes in their own right. The paradigmatic text presenting the new pure geometry is the *Elements* written by Euclid of Alexandria around 300 BC. In the geometric books of the *Elements*, one hardly finds any numbers at all. Theorems that in many other cultures, including our modern school of mathematics, are formulated in terms of numbers are formulated by Euclid in terms of pure geometry. For example, the first of the 13 books concludes with a proof of Pythagoras' theorem. In modern mathematics it is usually formulated by the equation $a^2 + b^2 = c^2$ where c is the hypotenuse and

[1] In fact, the unit 1 was not always considered a number by the Greeks.

a and b are the two other sides of a right-angled triangle. Here we think of a, b,and c as numbers measuring the length of the sides, and though we may say "the square of a" we just mean the number $a \cdot a$. For Euclid, on the other hand, the theorem is one about geometric squares. The square one can draw on the hypotenuse is just as big as the two squares drawn on the other two sides taken together.

Similarly in Book II, Euclid proves several theorems that we can translate into basic algebraic identities. For example, theorem 4 states the following: If a straight line is cut at random, the square on the whole equals the squares on the segments plus twice the rectangle contained by the segments (Figure 3.3).

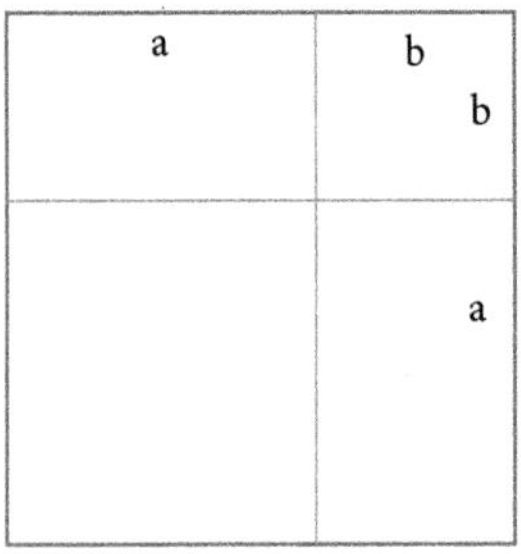

Figure 3.3 Illustration of Euclid's Book I Theorem 4 corresponding to the algebraic identity $(a + b)^2 = a^2 + b^2 + 2ab$

This theorem may be translated into the algebraic identity $(a + b)^2 = a^2 + b^2 + 2ab$. But where a modern reader will usually think of the formula as an identity involving numbers, the Euclidean theorem is a purely geometric theorem about squares and rectangles.

Ratios between geometric quantities posed a separate problem for the ancient Greeks. It was well known that the ratio of the corresponding sides a, a' and b, b' in two similar triangles is the same, i.e.,

$$a : a' = b : b'.$$

However, how should one interpret this identity when the sides are not measured by numbers, and $a : a'$ cannot be thought of as the number a divided by a'? The Greeks came up with at least two ingenious answers to this question, one of which was put forward by Eudoxos (*c.*408—*c.*355 BC), who worked at Plato's Academy in Athens. Eudoxos' pure theory of magnitudes was subsequently incorporated into Euclid's *Elements* as Book V.

Thus, the impossibility of finding a common measure of two arbitrary geometric magnitudes had a profound influence on the whole architecture of Greek mathematics. Moreover, since Greek mathematics in general and Euclid's *Elements* in particular became paradigmatic for much of the subsequent development of mathematics, the problem of incommensurability continued to shape mathematics well into the seventeenth century.

Historians of mathematics in the first half of the twentieth century often described the discovery of incommensurable quantities as a crisis in Greek mathematics. That may be so, but the feeling of crisis did not stifle the ancients. Instead of considering the discovery as a bar to progress, they considered it as a challenge. They soon found ways around the problem and thereby developed a great deal of highly original and important mathematics. As pointed out in the Introduction, this is the typical reaction to impossibility results. They rarely stifle mathematical progress. Rather they play a creative role, inducing mathematicians to invent ways to get around the impossibility.

The discovery of incommensurable line segments is the first example in the history of mathematics of a proof of an impossibility statement. Moreover, the method of proof (the indirect proof) became a paradigm for later impossibility proofs. In order to show that something is impossible you assume it is possible and reach a contradiction. This method worked quite easily in the proof of incommensurability, but it turned out to be much harder to obtain the contradiction in other cases.

3.3 Incommensurable quantities in Euclid's *Elements*

Given the revolutionary role played by the discovery of incommensurable line segments in the history of Greek mathematics, one would imagine that the existence of incommensurable magnitudes would occupy a central position in Euclid's *Elements*. However, that is not the case. To be sure, the indirect even–odd proof of the incommensurability of the side and the diagonal in a square is contained in some manuscripts of the *Elements* as theorem 117 of Book X. However, this theorem is generally considered a later interpolation. However, the definitions of Book X asserts the existence of incommensurable quantities. Having defined commensurability and incommensurability of two magnitudes, Euclid continued:

"**Definition 2.** Straight lines are commensurable in square when the squares on them are measured by the same area, and incommensurable in square when the squares on them cannot possibly have any area as a common measure.

Definition 3. With these hypotheses, it is proved that there exist straight lines infinite in multitude which are commensurable and incommensurable respectively, some in length only, and others in square also, with an assigned straight line … ." (Euclid 1956, Book X)

Such an existence theorem does not really belong among the definitions. Yet, Euclid probably felt that he needed to mention the existence of incommensurable quantities already here, so that the reader would know that the following definitions and the following theorems about commensurable and incommensurable quantities was not empty talk. He did something similar in the definitions of Book I: Having defined a diameter in a circle he added a theorem: "and such a straight line also bisects the circle." Here the reason is that he went on to define a semicircle as "the figure contained by the diameter and the circumference cut off by it." Obviously, it is nice for the reader to be told that indeed, the semicircle is half a circle.

Does Euclid then prove the existence of incommensurable line segments? In a sense he does in theorem 9 of Book X where he proves that two squares are to each other as a square number to a square number if and only if their sides are commensurable. From this theorem it follows that if we chose two geometric squares, whose ratio is not as a square number to a square number their sides are incommensurable (in fact this converse is formulated explicitly by Euclid). The question then is how one can find two squares whose ratio is not a square number to a square number. This problem is addressed in the proposition that follows. However, in some of the manuscripts of the *Elements*, this proposition has no number, indicating that it is a later interpolation. The wording of the proposition also supports this assumption. So apparently, Euclid did not quite deliver the promised proof. To be sure, it is quite easy to find two squares whose ratio is not a square number to a square number. Just take an arbitrary square and the square on its diagonal. They have the ratio 1:2, which is not a square number to a square number. Still, considering that Euclid is otherwise very meticulous, it is remarkable that he did not finish the argument himself.

This may indicate that although the Greeks realized how one could prove such impossibility results, they considered them as different from and perhaps less important than the positive results that one otherwise finds in the *Elements*.

4

Classical Problems of Antiquity

Constructions and Positive Theorems

The most famous impossibility theorems originating in ancient Greece are three classical construction problems. Just like the result about incommensurability, they were of great importance for the development of Greek mathematics. Knorr, in particular, has argued that these and similar construction problems were the driving force behind Greek mathematics. In his book, Knorr (1986) rejects the tendency among previous historians to depict Greek mathematics primarily as a theorematic endeavor, in which constructions of geometric problems served mostly as proofs of existence of the objects entering into the theorems. According to Knorr this is a distortion of the aim and lifeblood of Greek mathematics that can be traced back to late Greek philosopher-mathematicians who no longer understood the essence of the Greek mathematical project.

More traditional historians of Greek mathematics may not go as far but they all recognize the importance of the classical problems. However, the role played by these problems is different from the role played by the discovery of incommensurable quantities. Where the importance of the latter was primarily felt after the proof of impossibility had been found, the classical problems were mostly influential as problems crying out for solutions. In fact, as we shall see, the corresponding impossibility statements were late, slightly indirect, and without much consequence in Greek antiquity.

The three classical problems are the following:

1. The duplication of a cube: Given the side of a cube, construct the side of a cube with twice its volume.
2. The trisection of an angle: Given an angle, divide it into three equal parts.
3. The quadrature of a circle: Given a circle, construct a square with the same area.

As a necessary preliminary to the discussion of the impossibility question in Chapter 5, we shall in this chapter study some of the constructions of these

A History of Mathematical Impossibility. Jesper Lützen, Oxford University Press.
 DOI: 10.1093/oso/9780192867391.003.0004

problems that were put forward in antiquity as well as some positive results proved in this connection. For a more information the reader is referred to Heath (1921) and Knorr (1986).

4.1 Squaring a circle

The quadrature of a circle (or the squaring of a circle) is probably the most famous of all mathematical impossible problems. Indeed it is often used as a metaphor for (trying the) impossible. The problem goes back to Greek antiquity where it was already so popular that Aristophanes (*c.*400 BC) in his play *The Birds* could refer to it as a commonly known but exceedingly difficult problem.

Earlier cultures such as the Babylonians and the Egyptians had found approximate algorithms for calculating the area of a circle. For them, as for us today, the area was a number (of unit squares). Every schoolchild knows that the area of the circle with radius r can be calculated by the formula

$$A = \pi r^2,$$

where π is a number with infinitely many decimals beginning with 3.1415. However, such an answer to the question of the area of the circle was not an option for the Greeks. Indeed, as we saw in Chapter 3, the discovery of incommensurable quantities had made the Greeks aware of the fact that one cannot accurately measure all geometric quantities in terms of (natural) numbers. So instead of asking for a number measuring the area of a given geometric figure they asked for the construction of a square equal to it in area. This problem was called the quadrature of the figure.

It is interesting to note that the Egyptians had already found the area of a circle as a square whose side is $^1/_9$ smaller than the diameter of the circle. To be sure, this is not an exact method, but it is quite accurate, corresponding to a value of 3.16... for π. Thus, in a sense the Egyptians had an approximate solution to the quadrature of a circle. The Greeks, however, insisted on exact results.

In Book II of Euclid's *Elements*, one can find a construction of the quadrature of any polygon, i.e., any figure bounded by straight lines. This result probably goes back to the Pythagoreans. Let us see how one can square a rectangle.

Let $ABCD$ be a given rectangle (Figure 4.1). Extend the side AB beyond B and make $BE = BC$. Construct the midpoint of AE and draw a semicircle with

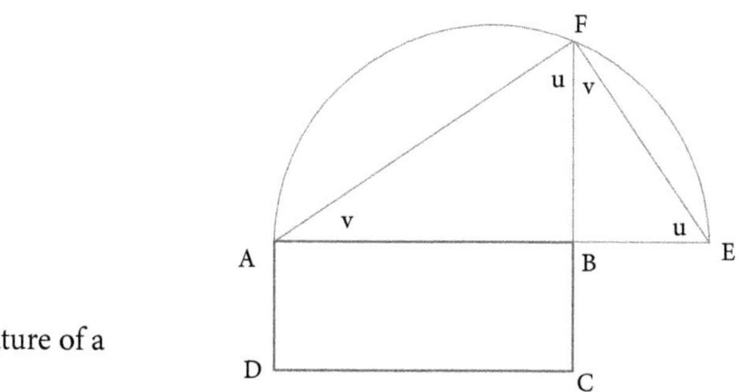

Figure 4.1 The quadrature of a rectangle

that center and *AE* as its diameter. Extend *CB* to its intersection point *F* with the circle. This completes the construction for we can show that the square on *BF* is actually equal to the given rectangle.

Indeed, if we draw the lines *AF* and *FE* the inscribed angle *AFE* is a right angle. This theorem has been ascribed to Thales. Since the sum of the angles in a triangle is equal to two right angles, and since the angles at *B* are right angles, the two other angles u, v in triangle *AFB* must add up to one right angle. But since the inscribed angle at *F* is also a right angle the angle *BFE* must also be v. And since the sum of the two angles at *F* and *E* in triangle *FBE* are equal to a right angle the angle at *E* must be equal to u. Thus, the two triangles *AFB* and *FEB* are similar and so the ratio of the corresponding sides are equal:

$$\frac{AB}{BF} = \frac{BF}{BE}. \tag{4.1}$$

This implies that

$$AB \cdot BE = BF^2.$$

Or as the Greeks would say: "The rectangle on *AB* and *BE* is equal to the square on *BF*." QED.

An arbitrary polygon can be divided into triangles that can easily be transformed into rectangles with the same area. By a nice little trick, Euclid could transform these rectangles into rectangles having the same height. They can be combined to one big rectangle that can be squared by the preceding construction.

Having thus found methods to square all polygons, the ancient Greeks turned to figures bounded by curved lines. Before we consider their investigations of this more complicated question, we shall pause to discuss the means used by the Greeks to solve construction problems.

It is usually said that Euclid made all his constructions by ruler and compass. In fact, Euclid never mentioned any instruments but his first postulates require that it is possible

1. to draw a straight line from any point to any point;
2. to produce a finite straight line continuously in a straight line; and
3. to describe a circle with any center and radius.

These constructions are precisely what one can do with ruler and compass. The above-mentioned quadrature of a rectangle and indeed all constructions in Euclid's *Elements* can be performed by repeated use of the three rules (postulate 1–3) and determining points of intersection between the drawn lines and circles. It is evident that Euclid and many other Greek mathematicians preferred such constructions by ruler and compass to other kinds of constructions, but it is up for debate who introduced this preference and when.

Many historians of Greek mathematics have suggested that it was Oenopides (mid-5th century BC) who formulated the preference for constructions using ruler and compass. The argument in favor of this attribution runs as follows: It is reported that Oenopides showed how to drop the perpendicular from a point onto a straight line. The only way this can be a memorable result is if Oenopides restricted himself to a limited number of instruments such as ruler and compass (but not including geometric triangles). However, Wilbur Knorr has cast doubt on this tenuous argument, maintaining that it is unlikely that such methodological ratings of various geometric methods we formulated so early (Knorr 1986, 15–17). Be that as it may, it is a fact that Oenopides's contemporary and associate Hippocrates of Chios only needed ruler and compass constructions for his quadrature of different "lunes." This beautiful result and argument is the first long piece of Greek mathematics that has been handed down to us more or less verbatim. I shall rephrase Hippocrates' proof slightly while keeping to the main idea:

A lune is a moon-shaped figure limited by two circular arcs. The simplest lune considered by Hippocrates is bounded by a half-circle *ACB* and a fourth of a larger circle *ADB* (Figure 4.2). Hippocrates found the midpoint *C* of the half-circle and drew the triangle *ACB* (note this can all be done by ruler and compass). He then claimed that the lune is equal to the triangle *ACB*. In order to prove this result Hippocrates noticed that the two circular segments I and II are segments in a quarter-circle just like segment III (see Figure 4.2). That means they are similar figures. Now Hippocrates knew that circles are to each other as the squares on their diameters and also that similar segments in circles are to each other as the squares on their chords. The first of these results is proved in Euclid's *Elements* XII.2. Euclid proves it using the method of exhaustion,[1] which is ascribed to Eudoxos. It is generally assumed that Hippocrates did not have a rigorous proof of the result, but the idea is intuitively plausible: If we draw two quarter circular segments in two circles of different size (Figure 4.3) and if the first of the segments has an area which is $k\%$ of the square on its chord, then the second circular sector is also $k\%$ of the square on its chord.[2]

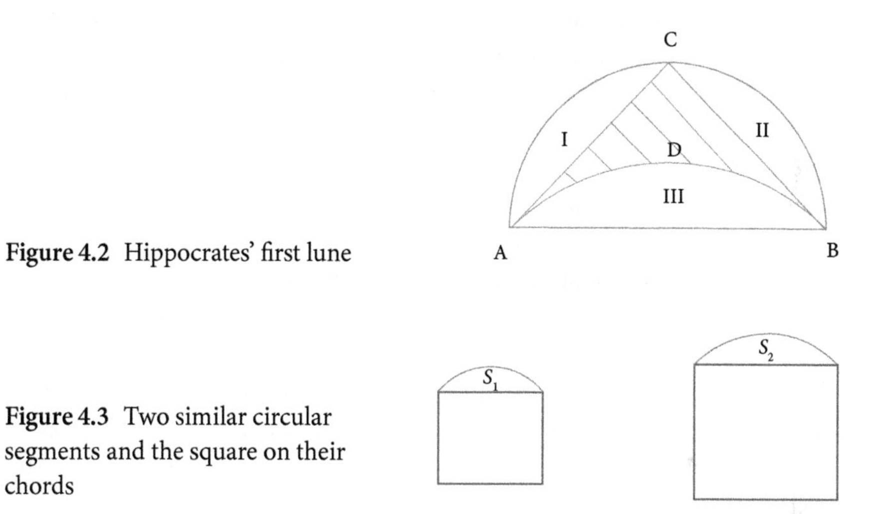

Figure 4.2 Hippocrates' first lune

Figure 4.3 Two similar circular segments and the square on their chords

Thus, the ratio of the two segments s_1/s_2 is the same as the ratio between the two squares.

Moreover Hippocrates observed that in triangle ABC (Figure 4.2), the angle at C is a right angle (Thales's theorem). Therefore according to Pythagoras's theorem [3]

$$AB^2 = AC^2 + CB^2.$$

If now we take $k\%$ of all the terms in this identity we find an identity between the three quarter segments in Figure 4.2:

$$III = I + II.$$

[1] We shall return to the method of exhaustion later.

[2] Of course the Greeks did not use %.

[3] I have used modern notation writing AB^2 where the Greeks would have said the square on the line segment AB.

Now add to each side the shaded area inside the triangle *ABC* lying above the quarter circle *ADB*. This leads to the desired result:

$$\textit{Triangle } ABC = \textit{Lune } ABC.$$

Since the triangle can be squared, so can the lune.

Hippocrates also squared two other lunes and showed that three equal lunes bounded by a half-circle and the sixth of a larger circle, together with the half-circle itself, is equal in area to a given trapezium. Later commentators such as Aristotle have interpreted this last result as an attempted circle quadrature. Indeed, if it was possible to square the last-mentioned lune, one would be able to square the (half-)circle because the trapezium is squarable. Since other lunes are squarable this gives hope of success. Whether this was really Hippocrates' strategy, it is certainly intriguing that some lunes can be squared while the seemingly simpler circle itself evades all attempts.

Most of the later Greek discussions of the quadrature of a circle were connected to the *rectification* of the circle. This problem calls for the construction of a line segment equal in length to the circumference of a given circle. The connection between the two problems was established by Archimedes (*c.*287–212 BC). Indeed, in his *Measurement of the Circle* he proved (Figure 4.4) that a circle has the same area as a right-angled triangle with one side adjacent to the right angle equal to the radius of the circle and the other side equal to the circumference.

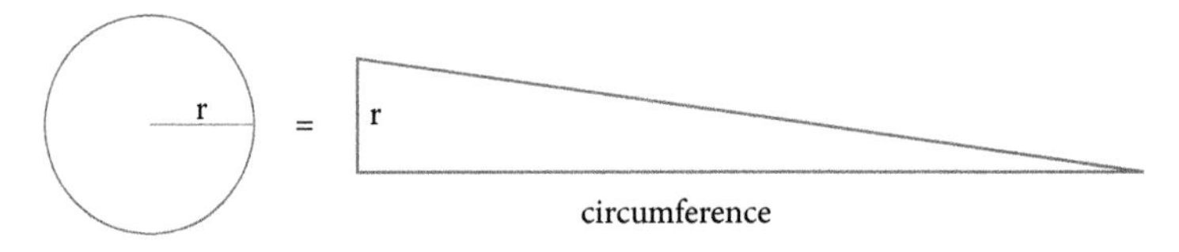

Figure 4.4 Archimedes' result connecting the quadrature and the rectification of a circle

The proof given by Archimedes is a double indirect proof by exhaustion. He first assumed that the circle is smaller than the mentioned triangle. In that case he could find a regular polygon circumscribed around the circle which is smaller than the triangle. However, he could easily show that a circumscribed polygon is larger than the triangle. This is a contradiction, and thus he could conclude that the circle cannot be smaller than the triangle. Similarly, by inscribing polygons in the circle he could show that the circle cannot be larger than the triangle. Thus, the circle must be equal to the triangle.

Since the triangle can be squared, this implies that one can square the circle if and only if one can rectify its circumference. Moreover, in a work on the sphere and the cylinder, Archimedes gave exact proofs of geometric theorems corresponding to our formulas for the volume of the sphere and the cylinder as well as the surface of the sphere. It follows from these results that if one can rectify the circle one can determine the area of the circle and the surface of the sphere as well as the volume of the cylinder, the cone, and the sphere. In modern terms these all depend on one constant, namely the ratio between the circumference and the diameter of the circle, i.e., the number that would later be called π.

Archimedes himself calculated (with great diligence) an upper bound for the circumference of a 96-gon circumscribed around the circle and a lower bound for the inscribed 96-gon. In this way, he established that the said constant (i.e., π) satisfies the inequalities

$$3\frac{10}{71} < \frac{\textit{circumference}}{\textit{diameter}} < \frac{22}{7} = 3\frac{10}{70}.$$

This gives narrow limits, but what about exact constructions? In fact, two constructions of the rectification of the circle have come down to us from Greek antiquity. One comes from Archimedes himself and uses the Archimedean spiral. This curve is generated by a point P that moves uniformly on a straight line while the line rotates uniformly around P's starting point O (Figure 4.5). Now consider the point R that the moving point P has reached after one full turn of the rotating line. The tangent to the spiral in R will cut the orthogonal OB to OR in a point B such that OB is equal to the circumference of the circle with radius OR.

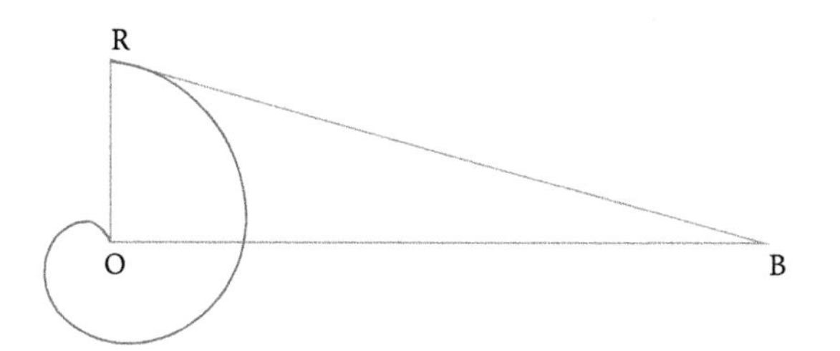

Figure 4.5 The Archimedean spiral and its tangent after one re-volution. *OB* is equal to the circumference of the circle with radius *OR*

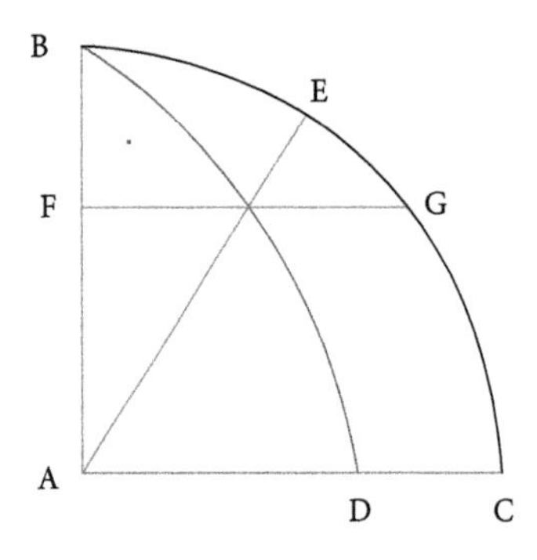

Figure 4.6 The curve *BD* is a quadratrix

The other construction of the rectification of the circle uses a curve aptly called the quadratrix (Figure 4.6). It is generated by the points of intersection of two moving straight lines *FG* and *AE*. *FG* moves uniformly downward parallel to itself and *AE* rotates uniformly around *A*. When they start, *F* and *E* are both in *B* and the motions are coordinated in such a way that they end up along *AC* at the same time. Now one can show that the point D where the quadratrix meets AC can be used to rectify the circle. Indeed

$$\frac{\textit{quarter arc BGC}}{AB} = \frac{AB}{AD}.$$

If the radius *AB* and the point *D* are given, one can easily construct a line segment equal to the quarter arc *BGC* satisfying this relation and thus a line segment equal to the entire circumference.

According to the ancient sources, three persons play a role in the history of the quadratrix: Hippias (perhaps Hippias of Elis, fl. 420 BC), Dinostratus (*c.*390–*c.*320 BC) and Nicomedes (*c.*280–*c.*210 BC). However, the sources and the modern experts do not agree on their respective roles. Traditionally, Hippias has been accredited with inventing the curve as a means to trisect an angle, Dinostratus has been given credit for its use in the quadrature of a circle, and Nicomedes for giving the exact Archimedean-style proof. Knorr, on the other hand, has argued that Hippias of Elis had nothing to do with the quadrarix, Dinostratus used it to trisect angles, and Nicomedes discovered how it could be used to square the circle, and therefore gave the curve its current name.

The merit of the construction of the rectification of the circle by way of the spiral or the quadratrix was debated already in antiquity. First, although Archimedes' result about the tangent to a spiral has often been interpreted as a rectification of the circle, it is unlikely that Archimedes himself thought of it in that way. Indeed, the problem of drawing a tangent to the spiral appears as a more complicated problem than the circle rectification itself. Therefore, it is more likely that conversely, he considered his result as a way to construct tangents to his spiral, given the rectification of the circle.[4] This would put this result on a level with his earlier mentioned results concerning volumes and areas of circles and spheres.

The argument using the quadratrix was explicitly presented as a solution of the rectification (and thus the quadrature) of a circle. However, it too suffers from a major problem that also complicates Nicomedes' proof: when the two moving lines reach the bottom position *AC*, their intersection point becomes undetermined. That means that the crucial point *D* must be defined as the limit of the intersection points as the two lines approach *AC*. By successive halving of the segment *AB* and the right angle *BAC* one can construct as many points on the quadratrix as one wants by ruler and compass. They can get arbitrarily close to *D*, but the point *D* itself cannot be constructed by ruler and compass.

Another critique that was raised against both curves was that somehow the rectification was already build into the curves through their coordination of the rectilinear motion and the circular motion. Does the use of the spiral and the quadratrix beg the question?

4.2 Doubling the cube

The problem of the duplication of a cube is also called *the Delian problem*. Here is why:

> For Eratosthenes says in his writing the *Platonicus* that when the god pronounced to the Delians in the matter of deliverance from a plague that they construct an altar double of the one that existed, much bewilderment fell upon the builders, who sought how one was to make a solid double of a solid. Then there arrived men to enquire of this from Plato [*c*.425–348 BC]. But he said to them that not for the want of a double altar did the god prophesy this to the Delians, but to accuse and reproach the Greeks for neglecting mathematics and making little of geometry.
>
> (Knorr 1986, 21)

The story of the oracle's instruction to the inhabitants of the island of Delos is told in several sources. Here it is Theon of Smyrna (fl. AD 100) quoting Eratosthenes of Alexandria (276–194 BC). The story is generally believed to be a legend, and it is certainly not an account of the origin of the problem.

[4] In fact, Archimedes constructed the tangent to the spiral in all of its points. That would be unnecessary if he wanted to use the tangent construction to determine the circumference of the circle.

Indeed many of the accounts of the legend including Eratosthenes' mention that the first advance in the solution of the problem was due to Hippocrates (the mathematician who squared the lunes) who lived half a century before Plato:

> It used to be sought by geometers how to double the given solid while maintaining its shape ... After they had all puzzled for a long time, Hippocrates of Chios was first to come up with the idea that if one could take two mean proportionals in continued proportion between two lines, of which the greater is double the smaller, then the cube will be doubled. Thus he turned one puzzle into another one, no less of a puzzle.
>
> (Eratosthenes, translated by Knorr 1986, 23)

Let us interpret the mathematics of this text in a somewhat modernized notation. Let a, b, x, and y be line segments. If

$$\frac{a}{x} = \frac{x}{b}, \tag{4.2}$$

we say that x is the mean proportional between a and b, and if

$$\frac{a}{x} = \frac{x}{y} = \frac{y}{b}, \tag{4.3}$$

x and y are said to be two mean proportionals (in continued proportion) between a and b. Now assume that we can construct the two mean proportionals x, y between the two given line segments a, b so that (4.3) holds. In that case we would have

$$\frac{a^3}{x^3} = \left(\frac{a}{x}\right)^3 = \frac{a}{x}\cdot\frac{a}{x}\cdot\frac{a}{x} = \frac{a}{x}\cdot\frac{x}{y}\cdot\frac{y}{b} = \frac{a}{b}. \tag{4.4}$$

Multiplying across[5] we get

$$a^3 b = x^3 a \ \ or \ \ a^2 b = x^3.$$

In the special case, mentioned by Eratosthenes where $b = 2a$ we have

$$2a^3 = x^3.$$

[5] Here I really sin against proper Greek thought according to which a four-dimensional object has no meaning.

Thus, x is the side of a cube which is twice as big as the cube with side a. So Hippocrates showed that if one can find two mean proportionals between two given line segments one can also double the cube. At first, this new problem may seem more tractable than the original problem. Indeed, it was known how to construct one mean proportional between two given line segments. This was exactly what we did (probably following the Pythagoreans) when we squared the rectangle (Figure 4.1 and Eq. (4.1)). Therefore, it sounds plausible that one can massage this construction to yield two mean proportionals. However, as remarked by Erathostenes some 200 years later the new problem of the two mean proportionals turned out to be no less of a puzzle than the original problem.

We saw that in the case of the quadrature of a circle, Eratosthenes had also "reduced" the problem to another problem, namely to the quadrature of a particular lune. But where that "reduction" led nowhere, the "reduction" of the Delian problem to the more general problem of finding two mean proportionals turned out to hold the key to all subsequent constructions of the problem.

Eratosthenes recounts that three mathematicians around Plato suggested constructions of the two means between two given line segments. Archytas (*c.*380 BC) came up with an ingenious stereometric construction based on the intersection of a cylinder, a cone, and a torus. It is too complicated to be explained here. Eudoxos is reported to have solved the problem by "curves" but no details of the method are known. On the other hand, Menaechmus' (*c.*350 BC) solution using conic sections has been preserved. In modernized notation it goes as follows: From (4.3) Menaechmus deduced that

$$x^2 = ay \text{ and } y^2 = bx \text{ and } xy = ab.$$

If we think of x and y as rectangular coordinates (which is almost what Menaechmus did), the points (x, y) satisfying the first equation lie on a parabola with axis along the y-axis, the points satisfying the second equation lie on a parabola with axis along the x-axis, and the points satisfying the last equation lie on a hyperbola with the x- and y-axes as asymptotes (Figure 4.7). The point (x, y) corresponding to the two mean proportionals must lie on all three curves. Thus, the mean proportionals x and y can be found as the x- and y- coordinates of the intersection between two of the three curves.

In this construction, Menaechmus assumes that one can draw parabolas and hyperbolas, just as Euclidean constructions assume that one can draw circles and straight lines. According to Eratosthenes, Menaechmus' construction

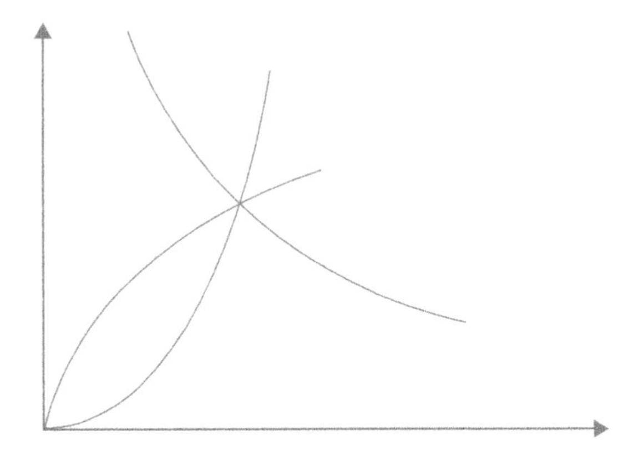

Figure 4.7 The triads of Menaechmus

relied on "sectioning off from the cone the triads of Menaechmus." This has often been read as saying that already Menaechmus defined the parabola and the hyperbola as plane sections of a cone. Since Eratosthenes speaks of a triad, it is often assumed that he also introduced the third kind of conic section, the ellipse. Moreover, since the triads are called after Menaechmus, it has been argued that the conic sections have their origin in his determination of the two mean proportionals. However, all of these interpretations have been questioned. The triads may be the two parabolas and the hyperbola used in Menaechmus' construction, he may have defined or described them in a different way, and the conic sections may have had another origin (see Knorr 1986). Nevertheless, we know for certain that with Euclid and Apollonius of Perga (*c*.250–*c*.190 BC) the parabola, the hyperbola, and the ellipse were defined as intersections between a cone and a plane.

Eutocius attributed another construction to Plato himself. It uses a mechanical device consisting of two rigid right-angled triangles that can slide in a specific way along one another. However, since Plato normally repudiated the use of mechanical devices in geometry, this attribution is generally rejected. Still, historians have suggested that Plato[6] could have used it to say: Here is an easy mechanical determination of the two mean proportionals, but of course, that is not what we are looking for. We want a proper geometric solution. Later Eratosthenes invented a similar mechanism, the so-called mesolabium, that could find any number of mean proportionals between two given line segments.

[6] Or at least Eutocius' account of Plato. See Knorr (1986, 57–9) and Van der Waerden (1961, 163–5).

Other post-Euclidean Greek mathematicians found other constructions of the two mean proportionals, using other instruments and other curves. Let me just mention the construction by Nicomedes whose circle quadrature using the quadratrix we met in the previous chapter. His determination of the two mean proportionals used another curve, the so-called conchoid (Figure 4.8).

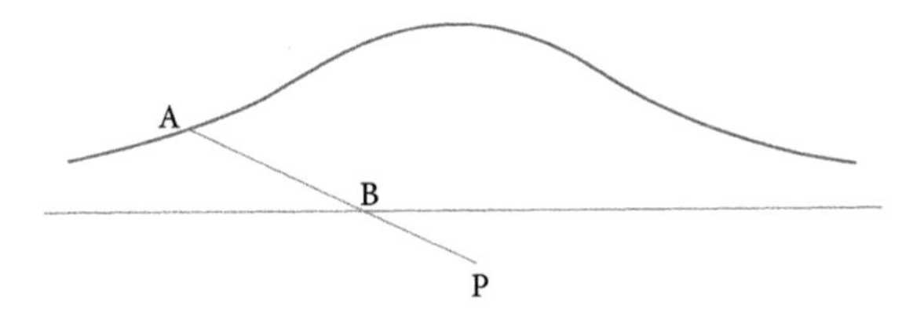

Figure 4.8 The conchoid of Nicomedes

The conchoid has the property that the line segment AB cut off between the curve and a fixed straight line is a constant if the line AB goes through a fixed pole P. I shall skip Nicomedes' intricate construction.

4.3 Trisecting the angle

There is no legend surrounding the trisection of an angle. However, the problem is quite natural. Indeed, it is easy to trisect a line segment (or even divide it into any number of equal parts) and to halve an angle by ruler and compass. So why should the trisection of an angle be more complicated? The earliest recorded method to trisect an angle uses the quadratrix. Through its coordination of a rectilinear and a circular motion, this curve is aptly designed to transform a trisection of an angle into a trisection of a line segment.

Indeed, given an angle CAE (smaller than a right angle), inscribe it in a quadratrix as seen in Figure 4.9. Draw the horizontal line FG and trisect the

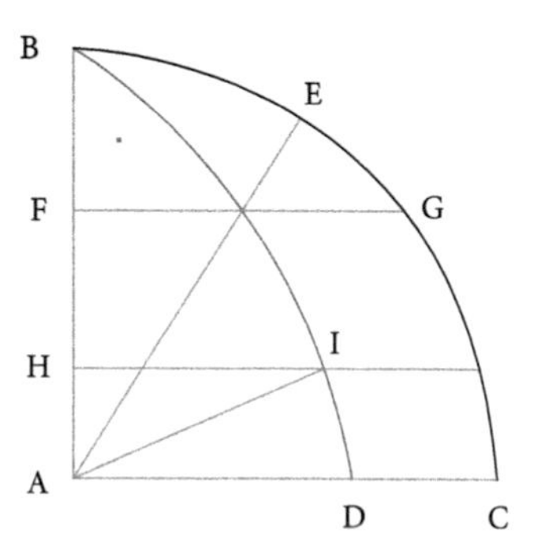

Figure 4.9 Angle-trisection using the quadratrix

line segment *AF* at *H*. Draw the horizontal line *HI* to its intersection I with the quadratrix. Then the angle *CAI* is a third of the original angle *CAE*. As mentioned in Section 4.1, this angle trisection has been attributed to either Hippias or Dinostratus.

Archimedes found a simpler construction. It runs as follows: Given an angle *AOB* (Figure 4.10), produce the leg *AO* beyond *O* and draw a semicircle *ABD* with center at *O*. Place a line *BC* going through the point *B* and meeting the circle and the line *AO* produced in two points *D* and *C*, such that *DC* is equal to the radius of the circle. Then the angle *ACB* is a third of the original angle *AOB*. To prove that, let v denote the angle *ACB*.

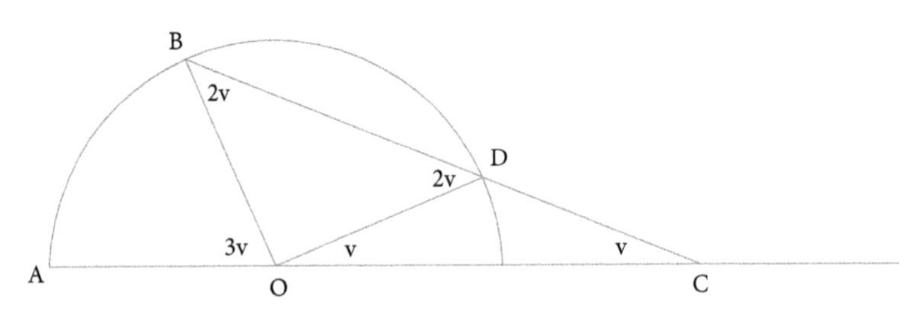

Figure 4.10 Archimedes' angle trisection

Since *DC* and *OD* are both equal to the radius of the circle, the triangle *ODC* is an isosceles triangle and so the angles at the base line are equal and both equal to v. The angle *ODB* is a supplementary angle to an angle in the triangle *ODC* and it is therefore equal to the sum of the remaining two angles in the triangles, i.e. equal to $2v$. The triangle *OBD* is also isosceles, so angle *OBD* is also equal to $2v$. Finally, the angle *AOB* is a supplementary angle to an angle in the triangle *OBC* and thus equal to the sum of the other two angles, i.e., equal to $3v$. Therefore, we conclude that the angle *ACB* is indeed a third of the original angle *AOB*.

The Archimedean construction is surprisingly simple and may at first sight look as a construction by ruler and compass. Indeed, we only used ruler and compass, or circles and straight lines in the construction. However, a closer inspection reveals that we have used the ruler in an unconventional way. According to Euclid's postulates the ruler can only be used to draw and prolong straight lines through two given points. However, when Archimedes drew the line *BDC* it went through one given point *B*, but the other two points *D* and *C* were not given. One has to rotate the line through *B* until the segment *DC* cut off between the extension of the line *AO* and the circle is equal to the radius of the circle. Such a construction is called a verging construction (because the line *DC* verges through *B*) or a *neusis* construction.

It is rather obvious that neusis constructions are linked with Nicomedes' conchoid. In fact any construction that can be performed with a neusis construction can be performed using a conchoid and vice versa. So it is not surprising that Nicomedes invented a trisection method using the conchoid.

In this chapter, we have seen how the three classical problems were solved in a number of ways using various curves or special instruments. That the Greeks continued to find new constructions bears witness to a feeling that the known constructions were not quite satisfactory. From the time of Euclid (300 BC) or a little later, Greek mathematicians strove for solutions using only ruler and compass. "And many and the greatest [of the geometers] sought, but did not find it" (Ammonius *c.* AD 440–520, quoted in Knorr 1986, 362). The unsuccessful search for such solutions gradually gave way to a conviction that the problems may simply be impossible with those means.

5

The Classical Problems

The Impossibility Question in Antiquity

5.1 Existence and constructability

Before we discuss what the ancient Greeks had to say about the impossibility of the classical problems, we shall analyze the distinction between existence and constructability. This distinction has been the object of some debate. Zeuthen (1896) argued that constructions functioned as existence proofs in ancient geometry. He based his claim primarily on the first books of Euclid's *Elements* where it is conspicuous that Euclid was very careful to construct all the geometric points, lines, and figures that he used in his arguments. However, it was noted long ago that Euclid in some of his later books does not follow this pattern. In particular Euclid has been criticized for assuming the existence of a fourth proportional to three surfaces in his proof of the theorem stating that two circles are to each other as the squares on their diameters (Euclid XII.2).[1] Where a construction of a fourth proportional to three lines is a simple matter, Euclid did not explain how to construct the fourth proportional when the given magnitudes are squares and circles. Indeed such a construction would require the theorem he was in the process of proving. Rather than accusing Euclid of making a mistake, Knorr and other historians have suggested that there were in Greek mathematics other means, in addition to construction, of establishing existence.

For example, Euclid may have argued that obviously, there exist circles that are too small to be the fourth proportional and there are circles that are too large. Therefore, there must by continuity exist a circle that is just the right size. Similarly, in the case of the quadrature of the circle one can argue as follows: It is obvious that the inscribed square is smaller than the circle while the circumscribed square is larger. Thus by varying the size of a square continually from that of the inscribed square to that of the circumscribed square, there

[1] Let a, b be quantities of the same kind and c a third given quantity, possibly of another kind. Then x is called a fourth proportional to a, b, and c if $a/b = c/x$.

A History of Mathematical Impossibility. Jesper Lützen, Oxford University Press.
 DOI: 10.1093/oso/9780192867391.003.0005

must, by continuity, be one intermediate square that is exactly equal to the circle. An argument along these lines was put forward by the sophist Bryson (fl. late fifth century BC) (Knorr 1986, 76–8).

Moreover, it seems obvious that construction was considered to yield more than mere existence. For example Euclid (III.1) shows how "to find the center of a given circle." Now, a circle is per definition "a plane figure contained by one line such that all the straight lines falling upon it from one point among those lying within the figure equal one another" (Euclid's *Elements* I, definition 15). Thus, if a circle is given, its center must exist. Thus, the construction cannot be intended to show the existence of the center, but must give more than mere existence.

Therefore, it seems to be in accordance with ancient thought to assume that the classical problems were not considered as problems of existence of the mean proportionals, the trisected angle, or the square equal to a given circle, but as problems of finding construction procedures of the said geometric objects.

There is, however, one exception: Aristotle actually stated that one cannot compare straight lines and curved lines. This statement, which seems to exclude the rectification of the circle even on the existential level, had a great influence on mathematicians well into the sixteenth century. It may seem as a strange claim, but Aristotle had good reasons for it. Indeed, even the later Euclidean *Elements* had no means of comparing the straight and the curved. In the *Elements* all comparisons of the magnitude of two quantities of the same kind are ultimately based on the axiom that the whole is greater than the part. However, given a curved and a straight segment, neither of them is a part of the other. To make up for that, Archimedes in the *Sphere and the Cylinder* formulated an axiom according to which the innermost of two convex curves with the same endpoints is the shorter. This is the key to the comparison of the straight and the curved.

Among post-Archimedean mathematicians and philosophers the existence of the figures sought for in classical problems were generally taken for granted. For example Philoponus (fl. *c.* AD 550) accepted Bryson's argument mentioned above as an existence proof, but denied that this was what the problem of the quadrature asked for:

> Those squaring the circle did not seek (to establish) whether it is possible that there exists a square equal to the circle, but rather thinking that it can be so, they tried to produce a square equal to the circle. (Philoponus, translated by Knorr, 1986, 77)

In the same vein, Eutocius (*c.* AD 480–540) wrote:

> For it is somehow clear to everyone that the circumference of the circle is some magnitude, I believe, and this is among those extended in one [sc. dimension], while the straight line is of the same kind. Even if it seemed not yet possible to produce a straight line equal to the circumference of the circle, nevertheless, the fact that there exists some straight line by nature equal to it is deemed by no one to be a matter for investigation. (Eutocius, translated by Knorr 1986, 362)

5.2 Pappus on the classification of geometric problems

The most thorough ancient account of the impossibility of the classical problems is Pappus's *Collection* from around AD 340. The context is Pappus's division of construction problems into plane, solid, and line-like problems. This division is discussed in Book III, Chapter VII and repeated almost verbatim in Book IV, Chapter XXXVI of the *Collection.* According to Pappus a problem is *plane* if it can be solved (i.e., constructed) by straight lines and circles (i.e., by ruler and compass), it is *solid* if it can be solved using the conic sections, and it is *line-like* if it can be solved by other more complicated curves (lines). He explains that the solid problems are so called because the conic sections are generated by the intersection of surfaces of solid figures, namely cones and planes. According to Pappus, the classification has normative aspects: He insisted that it is "no small error to geometers" if they solve a plane problem via conic sections or other more complicated curves and in general if a problem is solved by a method not belonging to its type (*Collectio,* Book III, Chapter XXXVI). Pappus attributed the classification to "the ancients" and indeed, the related division of loci can be found in Apollonius' work *On Planar Loci* (about 220 BC) and Aristaeus' earlier *On Solid Loci.* Yet, as mentioned earlier, there is no general agreement about the origin of the preference for ruler and compass constructions. To be sure, Euclid's *Elements* (about 300 BC) codified constructions by these means, but according to Knorr (1986, 346) it is not until Apollonius that we see a clear preference for ruler and compass constructions above other methods of construction. For example, Knorr points out that Hippocrates of Chios (second half of the fifth century) constructed his third lune by way of a neusis construction although it can be constructed by ruler and compass. The preference for solutions by conic sections over other types of construction is even less documented in the ancient literature. In fact, as Knorr argues, many ancient mathematicians seem to prefer neusis constructions and other "mechanical" constructions to solutions by conic sections. For example, we have seen that Menaechmus (mid-fourth century BC) found the two mean proportionals by the intersection of two parabolas or one parabola and one hyperbola. And yet, many of his successors invented many other solutions by way of other more complicated curves and by way of neusis or other mechanical means. In fact, Pappus himself is no exception. In addition to the solid construction of the two mean proportionals and the trisection of an angle, he reports many other constructions, one of which he claims for himself. So, even if there was an awareness of a classification of geometric problems before Pappus, it does not seem to have been sharp or even unanimously adopted, and it did not hold a strong normative status.

So much for Pappus's ban on constructing a problem with too complicated means. For our purpose, it is more important to consider the question: How does one rule out simpler constructions? More specifically, how does one know that a linear problem cannot be constructed by solid or plane means or that a solid problem cannot be constructed by plane means? Here Pappus leaves his readers in the dark. For example, he gives a nice construction of the trisection of an angle (or rather of a circular arc) by intersecting a circle and a hyperbola (*Collectio*, Book IV, Chapter XLIV) and then without further argument declares that he has shown that the problem is solid (Chapter XLV). Similarly, in the last-mentioned chapter he formulates the proposition that the division of an angle into a given ratio is line-like and promises two different proofs. However, what he offers is only constructions using the quadratrix and Archimedes's spiral, but no argument to support that the division cannot be done using ruler, compass, and conic sections. Such arguments would not be required if Pappus had considered the higher classes to include the lower ones, so that a plane problem would a fortiori be solid and line-like and a solid problem would also be line-like. However, this is clearly not his intention. He explicitly states that the solution of a solid problem "necessarily employs" conic sections. In addition, the normative aspects of the classification would lose its sense unless the problem classes are mutually exclusive.

Thus when Pappus declared that the problem of the two mean proportionals and the trisection of an angle are solid, there is no doubt that he implicitly claimed that they cannot be constructed by ruler and compass. But what kind of impossibility did he have in mind? Did he claim that the problems were in principle impossible or did he just make an empirical statement about his own and his predecessors' inability to solve the problems with these means?

He certainly appealed to the empirical fact that the ancient geometers could not find plane constructions of the two mean proportionals (Book III, Chapter VII) and the trisection of an angle (Book VI, Chapter XXXVI), but he also implied that there were deeper reasons for declaring the problems as impossible. Indeed, in both cases he explained that the ancients could not find plane solutions because the problems are solid *by nature*. Moreover, at the beginning of Book III Pappus rejected a plane construction of the two mean proportionals by one of his contemporaries who according to Pappus had a reputation as a great geometer. And not only did Pappus show that the proposed construction fails to solve the problem, he also ridiculed the unnamed geometer for trying to do the impossible.

Pappus's explicit categorization of the problems of the two mean proportionals and the trisection of an angle as being by nature solid and his resulting rejection of the possibility of their construction by ruler and compass are the closest any Greek mathematician came to a formulation of the impossibility theorems we are investigating. It is unclear what it is in the nature of the problems that made Pappus think they were impossible with plane means. He emphasized that it is important to "establish what is possible and what is impossible, and for that which is possible, when, how, and in how many ways it is possible" (*Collectio*, Book III, Introduction). However, Pappus never explains how one could establish the impossibility of the constructions. It is not even clear whether he understood the need for such a proof, or whether he considered such a result as provable at all. The only statement by Pappus that may be interpreted as an argument for the impossibility of the two mean proportionals is the following passage: "The ancient geometers could not construct the ... problem of the two straight lines, which is solid by nature, in a way conformable to geometric reasoning, because it is not easy to draw the conic sections in the plane" (Pappus, *Collectio*, Book III, Chapter VII). The corresponding statement about the trisection of an angle is subtly different. As for this problem Pappus stated that the ancients could not solve it by plane means "because the conic sections were not familiar enough to them" (Pappus, *Collectio*, Book VI, Chapter XXXVI). These arguments are difficult to follow. How did lack of knowledge of the conic sections hinder the ancients in finding a construction by ruler and compass? The only possible interpretation seems to be the following combination of the two arguments: If the ancients had been familiar with the conic sections, they could have found the constructions using these curves. If moreover one had been able to easily draw conic sections in the plane, these solutions would have passed for plane solutions. However, since the conic sections are difficult to draw in the plane, solutions using conic sections cannot be accepted as plane solutions. Even interpreted in this favorable way, the argument cannot pass as a proof, and clearly Pappus did not intend it to have the status of a geometric proof.

5.3 The quadrature of a circle

As pointed out by Knorr, Pappus's treatment of the quadrature of a circle was subtly different from his treatment of the two other classical problems. Where he was convinced of the impossibility of constructing the angle trisection and the two mean proportionals by ruler and compass, and thus convinced that the simplest possible solutions, namely those by conic sections had been found, he did not make such a claim concerning the quadrature of a circle. He showed that the rectification of the circle, and thus according to Archimedes also its quadrature, can be found by way of the quadratrix, but he did not state that the problem was linear. Thus, he did not exclude a plane or a solid construction. Knorr also quotes Eutocius, Ammonius, and Marinus (late fifth century) for saying that the rectification and the quadrature of a circle are not yet constructed but might still be constructible (Knorr 1986, 362–3). This seems to mean that they did not think that the constructions using complicated lines such as the spiral and the quadratrix were satisfactory and did not exclude the existence of simpler solutions, perhaps solutions by ruler and compass or by conic sections. Also the late neoplatonist Simplicius (about AD 490–560) emphasized the open-endedness of the search for a solution of the quadrature of a circle.

> The reason why one still investigates the quadrature of the circle and the question as to whether there is a line equal to the circumference, despite their having remained entirely unsolved up to now, is the fact that no one has found out that these are impossible either, in contrast with the incommensurability of the diameter and the side (of a square). (Simplicius *Physica*. Quoted from Knorr 1986, 364)

Simplisius' statement about the circle quadrature reveals a much deeper understanding of the status of the problem than Pappus's impossibility statements about the angle trisection and the two mean proportionals. Simplicius clearly distinguished between not having found a construction and the impossibility of finding one. Moreover, he saw the need for a proof of the latter and understood that such a proof was wanting. His comparison with the proof of the irrationality of the side and diagonal in a square even indicates that he considered a proof of impossibility as being within reach of geometry. This is the only place I know of where a Greek mathematician or philosopher points to the desirability and possibility of a geometric proof of the impossibility of one of the classical problems. To be sure, he did not specify exactly which means of constructions he deemed permitted, nor did he reveal a strategy for such a proof of impossibility, but that does not diminish his farsightedness.

It may be surprising that it was the circle rectification that was seen as analogous to the incommensurability question of side and diagonal in a square. After all, the cube duplication has a much closer affiliation to that problem. As even a slave boy can recall (with a bit of help from Socrates) the diagonal is after all the side in the doubled square. Moreover, it is also easy to see that the usual even–odd argument for the incommensurability of the side and diagonal of the square can be applied almost verbatim to show the incommensurability of the side of a cube and the side of another cube of twice its content.[2] However, since an irrational line segment may very well be constructible, as is the case with the diagonal of a square, the irrationality of the side of the doubled cube does not enlighten the question of its constructability. This may well be the reason why this analogy was not highlighted.

In the case of the rectification of the circle, the Greeks had not even been able to establish the commensurability or incommensurability of the diameter and the circumference. Therefore, a solution of this question would be a worthwhile result, and can be seen as a first step to the solution of the constructability. Indeed, if the diameter and circumference had turned out to be commensurable, the rectification would have been constructible by ruler and compass.

5.4 Using non-constructible quantities: Archimedes and Ptolemy

Greek mathematicians never stopped searching for better solutions to the three classical problems and were apparently never quite satisfied with the known constructions. How did that affect their mathematical practice in general? Here I shall briefly mention three examples, two from Archimedes and one from Ptolemy where the lack of suitable constructions of the classical problems had an influence on their approach to other geometric questions.

[2] The irrationality of $\sqrt[3]{2}$ in modern terminology.

It is quite clear that Archimedes was keenly aware that no generally accepted construction of the rectification of the circle was known. Yet, in the *Measurement of the Circle*, Proposition 1, he started out by assuming a straight line set out equal to the circumference of the circle. Here the mere existence of the line equal to the circumference was enough for Archimedes, just as the existence of a fourth proportional was enough for Euclid in Book XII of the *Elements*.

However, just as Euclid in the first books of the *Elements* was careful to construct all the line segments and figures he used in his deductions, Archimedes also showed a keen awareness of the question of constructability in other places of his oeuvre. The most conspicuous example is his proof of Proposition 34 in his paper *On the Sphere and the Cylinder I*, stating that every sphere is four times bigger than a cone having a base equal to the greatest circle in the sphere and height equal to the radius of the sphere. At one place of his long double indirect proof he chose two line segments b and c lying between two given line segments a, d, such that the segments a, b, c, d are in arithmetic progression (the successive differences between them is the same). If he had chosen b and c such that they were mean proportionals between a and d (so that a, b, c, d would be in geometric progression) the proof would have been quite straightforward. Choosing an arithmetical progression complicates the proof and even left a hole in the argument that Eutocius felt he needed to fill in in his commentary to Archimedes' book. There is no doubt that Archimedes knew that the proof would have been simpler if he had chosen a geometric progression. That he still used an arithmetic progression instead must be because he knew that such a progression can easily be constructed by ruler and compass (by trisecting the difference between d and a), whereas the construction of a geometric progression or the two mean proportionals requires more complicated means. This is the more remarkable because Archimedes was not involved in constructing a geometrical problem but only needed the means as a part of an indirect proof.

That does not mean that Archimedes never used the existence of two mean proportionals. Indeed, in the second part of *The Sphere and the Cylinder* (propositions 1 and 5) (Archimedes 1897, lxvii) he "assumed" the construction of two mean proportionals. His attitude seems to have been that if he could do without two mean proportionals, he would avoid them, even at the cost of complicating the argument, but if he could not avoid them (as is the case in the two propositions in Book II) he would assume their existence.

Another example of the awareness of the unsatisfactory status of the classical problems can be found in Ptolemy's *Almagest* (*c.* AD 150), more specifically in his construction of his table of chords. This is the first surviving trigonometric table. In this part of applied mathematics, Ptolemy did not shy away from calculating approximate values for the length of line segments. Having derived a series of theorems about triangles and quadrangles inscribed in a circle, he calculated good approximations of their sides. The calculations used Pythagoras' theorem and thus his calculations involved the extraction of square roots, as well as addition, subtraction, multiplication, and division. In this way Ptolemy found the cord of an angle of 12° (in a circle with radius 60) and through successive halving of the angle he arrived at a value for the cord of 1½°. His aim was to construct a table of cords in steps of ½°, so he was faced with the problem of finding the chord of ½° from the chord of 1½°. However, he remarked:

> Now, if a chord, e.g. the chord of 1½° is given, the cord corresponding to an arc which is one-third of the previous one cannot be found by geometric methods. (If this were possible, we should immediately have the cord of 1°). (Ptolemy, translated in Toomer 1998, 54)

This is a very explicit impossibility statement, even more explicit than Pappus's later statements. But exactly what was it that Ptolemy claimed to be impossible, and in which sense did he mean that it was impossible? It seems rather certain that "geometrical means" refer to constructions by ruler and compass, so he most certainly claimed that angle trisection is impossible with these means. But it is not possible from the short remark to infer whether Ptolemy thought it was impossible in the sense that no such construction had been found, or if he intended to state that it is in principle impossible, such as Pappus seems to have thought. If Ptolemy intended the latter, he gave (as Pappus) no reason why it should be impossible.

However, in contrast to Pappus, Ptolemy hints at an interesting link between geometric constructions and numerical procedures for calculating the length of line segments. He stated in the quote above that if a line segment can be constructed by ruler and compass, then its length can be calculated. Given that Ptolemy determined the length of his line segments by extracting square roots, one might infer from the quote that Ptolemy was aware of the fact that if a construction is done by ruler and compass then the constructed line segments can be found by square roots. This would be a remarkable insight that plays an important role in the modern proof of impossibility. However, if such an insight should be attributed to Ptolemy, one must admit that it is expressed rather indirectly. Moreover, Ptolemy used the connection between geometric construction and the numerical calculation in a way different from the modern impossibility proof. In the modern proof we first prove that the chord of the trisected angle cannot (in general) be found by square roots and from there conclude that the trisection cannot be done by ruler and compass. Ptolemy, on the other hand, concluded conversely that since the geometric construction is impossible he could not calculate the chord of $1/2°$ (by square roots). Consequently, he approximated the chord of 1° by linear interpolation between the chord of $1\frac{1}{2}°$ and the chord of $3/4°$.

6
Diorisms
Conclusions about the Greeks and Medieval Arabs

6.1 Diorisms

In this chapter we shall study a class of problems in Greek geometry that sometimes have solutions and sometimes not, depending on the magnitudes given in the problem. In such cases a *diorism* is a condition involving the given magnitudes, such that there exists a solution if the condition is fulfilled but there does not exist a solution if it is not fulfilled. In other words, the diorism separates possible from impossible instances of the problem.

The easiest example is found in Euclid's *Elements* I, 22:

> To construct a triangle out of three straight lines which equal three given straight lines: thus it is necessary that the sum of any two of the straight lines should be greater than the remaining one.

The last condition on the three given lines is the diorism. In the construction of the triangle, Euclid begins by assuming that the sum of any two of the straight lines is greater than the third line. Then he gives a construction involving the intersection of two circles. The construction works because under the conditions on the given lines the two circles actually meet. However, Euclid never explicitly mentions where the conditions are used, assuming apparently that the reader can verify the existence of the intersection point by inspection of the figure. The axioms are not strong enough to guarantee the existence on their own. Despite these shortcomings, Euclid's intention is clearly to show that the conditions on the lines are sufficient to guarantee a solution. The necessity mentioned in the diorism is contained in proposition I.20:

> In any triangle the sum of any two sides is greater than the remaining one.

This theorem is the contraposition of the impossibility part of the diorism in proposition 22: The problem of proposition 22 is impossible unless the sums of any two of the line segments are greater than the remaining one.

A History of Mathematical Impossibility. Jesper Lützen, Oxford University Press.
 DOI: 10.1093/oso/9780192867391.003.0006

The combination of the two theorems show that the construction problem is possible if and only if the conditions mentioned in the diorism are satisfied.

A more complex problem involving a diorism can be found in Euclid's Book VI, proposition 28.

> To apply a parallelogram equal to a given rectilinear figure to a given straight line but falling short by a parallelogram similar to a given one; thus the given rectilinear figure must not be greater than the parallelogram described on the half of the straight line and similar to the given parallelogram.

This so-called elliptic application of areas is quite hard to understand for a modern reader. Let me explain it in the special case where the given parallelogram is a square. In that case, we have been given a line segment b and a polygon (a rectilinear figure) whose area c is the only important information. We are required to construct a rectangle with area c and sides x and y, such that when it is placed along (applied to) the line segment b there is missing (it is falling short by) a square (with side x) (Figure 6.1). Since the given line b is thus equal to the sum of the sides of the rectangle we can formulate the problem as follows.

Figure 6.1 The elliptic application of areas

Construct a rectangle of given area c, such that the sum of the sides is equal to a given line b. Or expressed in modern language: Solve the equations

$$x + y = b, \tag{6.1}$$

$$x \cdot y = c. \tag{6.2}$$

Recall that the Babylonians had already solved such a system of equations. It can be transformed into one quadratic equation in x:

$$x^2 + c = bx. \tag{6.3}$$

Euclid begins his construction by dividing the given line segment b into two halves. Then he constructs a square on each of the halves. In the upper left-hand corner of the rightmost of the two squares he places a smaller square (the gray square; see Figure 6.2) equal in area to that of the square on half of b minus the given area c:

$$\textit{grey square} = \left(\frac{b}{2}\right)^2 - c.$$

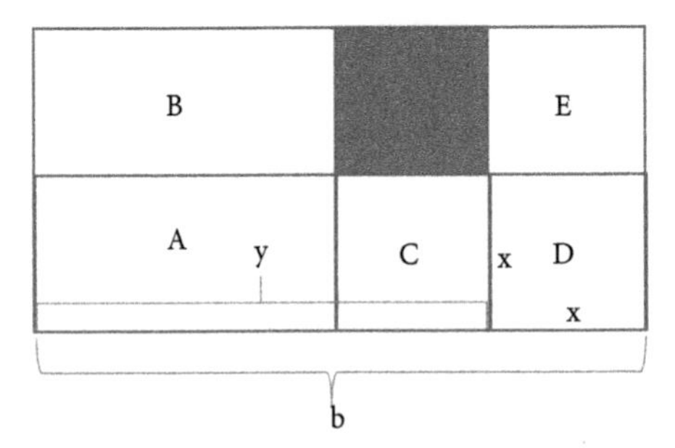

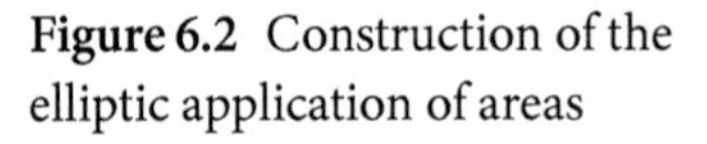

Figure 6.2 Construction of the elliptic application of areas

Then he completes the figure. It is now simple to see that the union of the rectangles A and C will constitute a rectangle with sides x and y that satisfy the requirements. Indeed, the "gnomon" consisting of the rectangles C, D, and E is equal to c because

$$C + D + E = \left(\frac{b}{2}\right)^2 - \textit{grey square} = \left(\frac{b}{2}\right)^2 - \left(\left(\frac{b}{2}\right)^2 - c\right) = c.$$

If the right-hand rectangle (D+E) is rotated 90° it will just fit onto A (both have sides $b/2$ and x) and so the area of the rectangle A+C equals C+D+E or c. Moreover, since D is a square it is clear that the sum of the sides of the rectangle A+C adds up to b. QED.

This construction works when the given area c is smaller than the square on half of the given line b, or in modern notation if $c \leq \left(\frac{b}{2}\right)^2$. Indeed, in that case, $\left(\frac{b}{2}\right)^2 - c$ is a non-negative polygon, and Euclid has shown how to construct (by ruler and compass) a square (the gray square) equal in area to it. The condition $c \leq \left(\frac{b}{2}\right)^2$ is precisely the diorism in the case we deal with here where the given polygon is a square.

In passing we remark, that Euclid's solution corresponds to the Babylonian algorithm or the modern formula for solving a quadratic equation. Indeed, since the gray rectangle has area $\left(\frac{b}{2}\right)^2 - c$, its side is $\sqrt{\left(\frac{b}{2}\right)^2 - c}$. Moreover it is obvious from the figure that x and y can be found by adding and subtracting this side to/from $\frac{b}{2}$, i.e.,

$$x, y = \frac{b}{2} \pm \sqrt{\left(\frac{b}{2}\right)^2 - c}.$$

This is exactly the formula for the solution of the quadratic equation (6.3).

So in the proof, Euclid showed how to solve the problem (by ruler and compass) when the diorism is fulfilled. In this problem too, he dealt with the necessity of the diorism in another proposition. Indeed, Euclid VI.27 states

> Of all the parallelograms applied to the same straight line falling short by parallelogrammic figures similar and similarly situated to that described on the half of the straight line, that parallelogram is greatest which is applied to the half of the straight line and is similar to the difference.

In the square case we deal with here, this theorem says that the largest rectangle one can apply to the line b, such that a square is missing, is the square on $b/2$. In other words if the area c is larger than the square on $b/2$ the elliptic application of areas is impossible. Or expressed in more algebraic terms: if $c > \left(\frac{b}{2}\right)^2$, the system of Eqs (6.1) and (6.2) or equivalently the quadratic equation (6.3) has no solutions.

This was precisely the impossibility result that we looked for but did not find in the Babylonian texts dealing with such equations. Euclid formulated the result and proved it. Still, we must admit that the necessity of the condition was formulated in a positive way, by contraposing the impossibility statement.

In the above analysis, I have translated Euclid's text into later algebraic notation. This was important for three reasons: 1. In order to explain the problem of the elliptic application of areas to a modern reader for whom the original formulation is difficult. 2. In order to compare Euclid's exact formulation and proof of the diorism with the situation in Babylonian sources where such considerations about possible and impossible cases are absent. 3. Finally, we shall return to the system (6.1) and (6.2) or the corresponding quadratic equation when we reach the Renaissance where the equivalence of these equations and Euclid's application of areas was well understood. Whether it makes sense to ascribe a kind of geometric algebra to the Greeks is a hotly debated subject I shall not deal with (see, e.g., Blåsjö 2016).

6.2 Conclusion on impossibilities in Greek mathematics

We have seen that the ancient Greek mathematicians dealt with impossibility statements in three different ways.

1. The impossibility parts of the constructions having diorisms were included as proper theorems with proofs in Euclid's *Elements*. But they were formulated as positive theorems. Instead of saying that the problem was impossible when the diorism was not satisfied, Euclid proved the contraposed statement to the effect that if the construction was possible, the diorism was satisfied.
2. The impossibility of finding a common measure of the side and diagonal of a square (or a regular pentagon) (the incommensurability question) was apparently the first impossibility theorem that was given a proof. It had a decisive influence on the development of Greek mathematics and became widely known. One of the possible proofs was a proof by contradiction. However, Euclid does not seem to have included this proof into the *Elements* although he used indirect proofs extensively elsewhere in the *Elements*.
3. Finally, the three classical problems were never proved impossible with ruler and compass although at least some late Greek mathematicians entertained the idea that they might be impossible with those means. Several mathematicians such as Pappus and Hero were convinced that the duplication of a cube and the trisection of an angle were by nature impossible with ruler and compass, but did not supply any mathematical arguments in favor of the impossibility.

Let us compare the three different types of impossibilities, starting with the different treatment of the diorisms and the incommensurability. Why did Euclid deal more directly with the diorisms than the more revolutionary incommensurability question? The answer may lie in the more negative nature of the incommensurability proof than in the diorisms. The proof of incommensurability is entirely negative. The problem of finding a common measure is proved to be entirely impossible. In theorems with a diorism there are possible cases and impossible cases. Euclid, and other Greek mathematicians, may have thought of themselves as problem solvers and thus thought more highly of theorems that would solve problems and explain when it can be done than theorems that proved the impossibility of a problem. To a problem solver an impossible problem may just seem as a bad problem. I admit that we cannot know for sure whether the Greeks thought along these lines, but it is at least consistent with the sources.

Now, let us compare the first two examples mentioned earlier with the classical construction problems. Why did the Greeks prove the impossibility of the former whereas they never proved the impossibility of the classical problems? The most immediate answer is: the impossibility of the classical problems is much harder to prove than the incommensurability and the diorism theorems mentioned in Euclid's *Elements*. The proof of impossibility of the classical problems requires methods that were not available to the Greeks.

This mathematical argument certainly explains why the Greek geometers did not come up with proofs of impossibility of the classical problems. However, it does not explain why one can only find one statement of the desirability of a proof, and hardly any awareness that such a proof was wanting or even possible. Given the knowledge of other proofs of impossibility, why did the impossibility of the classical problems not present themselves to the Greeks as a question that called for a proof? It is conceivable that the ancients realized, or just felt, that the impossibility of the classical problems is of a different nature than the incommensurability question and the diorism statements. In fact, in the last-mentioned problems, it is an object that does not exist: The common measure in the case of incommensurability and the figure that we are asked to construct in the case where the diorism does not hold. On the other hand, as we saw earlier, most Greek mathematicians and philosophers agreed that the objects that the classical problems ask for exist. The problem is not one of existence of an object, but one of the existence of a particular kind of construction procedure.

This makes it harder to imagine how one should even approach the impossibility statement with mathematical methods. For example, it seems hopeless to find an indirect proof. For such a proof to work, one should assume that a construction by ruler and compass exists and one should then deduce a contradiction. But that would require that one knew of properties belonging to constructible figures but not to the figures that the classical problems call for. However, the Greeks only had an intentional definition of constructible figures, not an extensional one. As long as such an extensional characterization is missing, such a property is hard to find.

The Greeks may also have contemplated whether a method of proving incommensurability using Euclid's algorithm might be generalized to the classical problems. In both cases, one needs to prove that a certain step-by-step procedure does not lead to the goal in a finite number of steps. However, the great difference is that Euclid's algorithm is an automatic algorithm. The two given quantities predetermine all the steps in the algorithm. In the case of the construction of, e.g., an angle trisector, there is no such determined road to go. At each new step in a successive construction, one must make a choice of whether to draw a line or a circle and where to draw it. How should one prove that none of the infinitely many choice sequences would lead to the desired result?

We do not know whether any Greek mathematician ever attempted to prove the impossibility of the classical problems or whether the difference between the non-existence of an object and the non-existence of a construction was so great psychologically or philosophically that the Greeks did not even contemplate the possibility of a proof; but the sources certainly indicate that the ancients dealt with the impossibility of the classical problems in a way different from the way they dealt with other mathematical problems.

Although there was a great interaction between Greek mathematics and philosophy, there was still a clear difference between mathematics and meta-mathematics. In mathematics one constructs figures and proves theorems; in meta-mathematics one discusses the philosophy behind this endeavor. Euclid, Archimedes, Apollonius, and the other mathematicians wrote treatises about mathematics, whereas Plato, Aristotle, and other philosophers wrote works about meta-mathematical questions. Only a few of the later philosopher-mathematicians wrote works that dealt with both subjects. Pappus' *Collection* is one such work. He discusses both methodological questions, such as the division of construction problems into plane, solid, and line-like problems, and he proved mathematical theorems. However, he treated the two types of questions in different ways. He proved mathematical results but did not provide proofs of meta-claims. For example, he did not try to prove that his classification of problems was correct. He gave arguments but no proofs. He dealt with the impossibility claims in the same way as any other meta-statement about mathematics.

It thus seems to me that Pappus and probably also most other Greek mathematicians thought of the impossibility of the classical problems as meta questions and therefore did not even contemplate that they could or should be proved.

6.3 Medieval Arabic contributions

During the Middle Ages the Arab,[1] Muslim culture continued the Greek (and Indian) mathematical tradition and developed it further. Algebra (an Arab word) was perfected to such a degree that it is fair to say that it is an Arabic science. Moreover, trigonometry was improved, and even in the field of geometry the Arabs continued many aspects of Greek mathematics. For example, they continued the tradition of geometric construction of problems. Although they do not seem to have known Pappus' *Collection*, they were familiar with and for the most part accepted his preference for solid constructions using conic sections rather than neusis and other mechanical constructions.

In particular, Arab mathematicians were proud of having succeeded in improving constructions they ascribed to Archimedes, whom they admired immensely. For example it was believed that Thābit ibn Qurra (second half of the ninth century) was the first to have found a solid method to trisect an angle. The Arabs knew that Archimedes had solved the problem using a neusis construction, so Thābit's construction was considered an improvement. However, as we have seen in Chapter 5. Pappus had already reported such a solid construction in his *Collection* and as pointed out by Hogendijk (1981), Thābit's construction is almost identical to Pappus'. Indeed Hogendijk has argued that Thābit knew the solid construction from a translation by Aḥmad ibn Shākir (before 803–73) of a Greek manuscript other than Pappus' *Collection.* The Greek origin of the method was later forgotten.

Concerning the possibility of trisecting an angle by ruler and compass, at least some Arab mathematicians believed it was impossible. For example, a commentary of al-Kāshī (*c.*1420) reported: "some of them said that there does not exist any method for knowing by means of lines[2] the chord of the third part of an arc whose chord is known" (Rosenfeld and Hogendijk 2002–3, 33).

[1] In this book the phrase "Arab mathematics" refers to mathematics written in the Arabic language during the Middle Ages.

[2] Meaning probably by ruler and compass.

However, no attempt of *proving* this impossibility has survived from the Arab medieval period.

The trisection problem mentioned here came up in connection with the construction of trigonometric tables. As we saw in Chapter 5, Ptolemy had struggled with finding the length of the chord of $1/2°$ from the length of the chord of $1 1/2°$ in a circle of radius 60. Many Arab mathematicians improved on Ptolemy's table, culminating with the tables constructed under supervision of the Samarqand ruler Ulūgh Beg (1394–1449) tabulating the sine for every minute of arc to 5 sexagesimal places. In this connection his subject al-Kāshī (about 1420) calculated $\sin(1°)$ to 9 sexagesimal places (or 17 decimals). He achieved this impressive accuracy by showing that $x = \sin(1°)$ is a root of the cubic equation

$$\sin(3°) = 3x - 4x^3,$$

and by inventing an ingenious iterative numerical method of solving this equation that he said he had "discovered by the power of inspiration from the Eternal Presence" (Rosenfeld and Hogendijk 2002–3, 46).

Al-Kāshī also determined π to 16 decimal places. As for the possibility of squaring a circle or finding an exact value of π the Arabs thought that it was known to God but not to us:

> There are questions of which the essence is possible in nature, but not for us, or the investigation of them is impossible for us because of the lack of preliminaries for them. Such is the quadrature of the circle. (al-Sijzī (10th century), quoted in Hogendijk 1996, 5)

The impossibility mentioned here is probably the impossibility of *constructing* the square equal to a given circle by a method which was considered legitimate. The existence of the square could be argued for by a continuity argument. For example, Ibn al-Haytham (*c.*1000) used a continuity argument in his study of conic sections (Hogendijk 1985, 95), and in his quadrature of a circle he similarly seems to have argued by continuity for the existence of the square equal to the given circle. He emphasized the difference between existence and constructability of the square in the following way:

> The essences of the knowable notions do not require that one perceive them or that they be actually produced. Nay, if the proof of the possibility of the

> notion has been perceived, the notion is sound, whether one has actually produced it or not. (Ibn al-Haytham, quoted from Hogendijk 1985, 96)

This echoes the remarks by Philoponus and Eutocius whom we met in Chapter 5.

The construction of the regular heptagon[3] was another problem that the Arabs studied intensely. Here they possessed a construction that they attributed to Archimedes.[4] This construction assumed a construction, similar to a neusis construction. However, instead of rotating a line around a point until the *distance* cut off by two given curves has a given length, it is here assumed that two triangles cut off by the line we seek are equal in *area*.[5] This construction is clearly more unacceptable than a normal neusis construction so it is not surprising that the Arabs rejected it as a proper construction. Instead, various Arab mathematicians from the tenth through the thirteenth century found 12 different correct constructions of the regular heptagon using only intersections of conic sections (Hogendijk 1984).

These constructions were indeed novel since the Greeks had not found any solid constructions of the regular heptagon. These and other additions to and improvements of Greek mathematics convinced some Arab mathematicians that they had finally reached a level of knowledge and inventiveness that equaled or even exceeded that of the ancients. Rosenthal (1950) quotes al-Asṭurlābī and al-Samaw'al (both mid-twelfth century) expressing this proud opinion. The latter wrote:

> Most of [the Arab mathematicians] assume that the ancients discovered all the knowledge that can be known; that nobody is able to know what they did not know; and that that which they did not know cannot be known, ... The contention that Euclid knew more about geometry than the many excellent scholars who lived before his time does not necessarily imply that Euclid might not be succeeded by someone who, like as Euclid was better than his predecessors, would be better than Euclid. There is, for instance, Archimedes. His book on the Sphere and the Prism[6] entitles him to such a rank. In his Lemmata, Archimedes now had to admit his inability to achieve the trisection of angles. Eventually, however, ... the trisection of angles (was achieved]

[3] A regular polygon with seven sides.
[4] Hogendijk (1984) has argued that the construction was probably Greek but not by Archimedes.
[5] See Hogendijk (1984) for more details.
[6] The work is rightly entitled *On the Sphere and Cylinder*.

> by ... Aṣ-Ṣâgânî and the construction of the heptagon in the circle by ... al-Kûhî ... . There still remain the division of angles into five equal parts; the construction of regular polygons of eleven, thirteen, and seventeen sides in the circle. Those problems are as yet unsolved, but it can be proven that a solution exists and is not impossible. The fact that their solution has been impossible for us and all our predecessors merely shows that the knowledge at hand and the available postulates are not sufficient to discover the solution and that other still unknown postulates are needed. It is not impossible that we will be succeeded by someone to whom God will show the solution. (Rosenthal 1950)

This self-assured and optimistic quote raises the question of impossibility that interests us in this book. It is clear that for al-Samaw'al, an impossible problem here means a problem that has not been solved at the indicated time. And indeed, in the literature concerning geometric constructibility, the Arab mathematicians did not discuss the possibility of proving that a problem cannot be constructed by certain means. However, al-Samaw'al did mention such impossibility proofs in his book *Al-Bāhir* that dealt with arithmetic algebra. In a methodological section towards the end of the book, he classified problems into three categories: The necessary, the possible, and the impossible problems.[7] The necessary problems are the problems where the existence of a solution can be demonstrated. "The impossible problems are those such that if you assume them existing[8] their existence leads to the impossible" (Rashed 2021, 239).[9] The possible problems are the ones where the mathematician can neither demonstrate the existence nor the nonexistence of a solution.

It is not quite clear to me how al-Samaw'al would classify "the division of angles into five equal parts; the construction of regular polygons of eleven, thirteen, and seventeen sides in the circle" that he mentioned in the previous quote. His optimism about a future solution seems to exclude that the problems should be classified as impossible, but are they necessary or possible? On the one hand, al-Samaw'al claimed that it can be proved that a solution of the problems exist. Here he probably had in mind a continuity argument similar to the continuity arguments used by Ibn al-Haytham. That would mean that the problems are necessary problems. On the other hand, if we ask for a solution by given means (e.g., by intersection of conic sections) the problems would be "possible."

[7] See Rashed (2021, 238–40).
[8] That is, "if you assume that a solution exists."
[9] That is, to a contradiction.

Al-Samaw'al divided the impossible problems into those that are impossible because of their definition and those that are impossible because of their hypotheses. As an example of the former he mentioned the problem of finding two natural numbers such that their ratio is as a square number to another square number, and such that their product is not a square number. Al-Samaw'al gave a brief argument to show that this is indeed impossible. His example of a problem that is impossible because of its hypotheses seems to indicate that he had in mind problems that are possible in some cases, and not in other cases. Here one needs a diorism to distinguish the necessary from the impossible cases.

The Arabs, in fact, found some quite remarkable diorisms both in geometry and in the theory of equations. As an example of the former, Berggren (1996) has discussed al-Kūhī's (second half of the tenth century) solid solution of a problem arising from Archimedes' book *The Sphere and the Cylinder*. The problem is the following: Construct a spherical segment whose (spherical) surface is equal to that of one segment and whose volume is equal to that of another. The problem is to determine the radius of the sphere and the height of the segment satisfying the requirements. Al-Kūhī constructed the solution by intersecting a hyperbola and a parabola. However, there is a problem: The problem may not always have a solution at all. The existence of a solution depends on the given values of the volume and the spherical area. This is also reflected in the construction procedure. Indeed, the parabola and the hyperbola may not intersect, and in case they do, it is not certain that both intersection points lead to a solution of the problem. By a very intricate argument, al-Kūhī was able to determine the condition (diorism) that had to be satisfied by the given volume and area, in order for the problem to have a solution.

Berggren speculates whether al-Kūhī might have known that the problem is equivalent to a cubic equation, but his surviving solution is entirely geometric. About a century later, Omar Khayyam was able to construct the solution of all types of cubic equations by intersecting conic sections. The corresponding question of the number of (positive) solutions was cleared up completely by Sharaf al-Dīn al-Ṭūsī (end of twelfth century). He showed under which conditions on the coefficients, there are zero, one, two, and three solutions of the equation.[10] His arguments are among the most sophisticated in medieval Arab mathematics, and have been claimed by Rashed to be a precursor of differential

[10] In Chapter 10 we will see that already around 800, al-Khwārizmi had cleared up the corresponding question for quadratic equations.

calculus. To be sure, Sharaf al-Dīn al-Ṭūsī used a concept similar to that of a function introduced in the West about 1700, but Hogendijk has cautioned that it is ahistorical to attribute infinitesimal ideas to Sharaf al-Dīn al-Ṭūsī (see Hogendijk 1989).

Be that as it may, Sharaf al-Dīn al-Ṭūsī's analysis of the solution of cubic equations is a remarkable instance of supplying diorisms to problems, and in particular of showing when they are impossible to solve.

7

Cube Duplication and Angle Trisection in the Seventeenth and Eighteenth Centuries

7.1 The seventeenth century

During the early modern period, the division between mathematics and meta-mathematics was less clear-cut than during the ancient Greek period. Many of the most important works of the seventeenth century had strong methodological elements. In particular, many of them dealt in some way or another with finite or infinitesimal analysis that was thought of as a method for solving problems and a method of discovery. In the analytic arts, there were arguments, but these were often of a rather intuitive and less rigorous sort than the ancient geometric proofs. One reason that such laxer standards were accepted was that the results obtained by analytic means were supposed to be followed by a synthetic proof. These synthetic proofs (when they were given) were still of the rigorous terse Greek type. In analysis, mathematics and meta-mathematics mixed more easily. This may be one of the explanations as to why impossibility theorems became more visible. They could exist comfortably in the analytic texts somewhere between meta-mathematics and mathematics.

However, Mancosu (1996) has pointed to a factor that may have made it more difficult for some seventeenth-century mathematicians to deal with impossibility theorems, namely the adherence to an Aristotelian distinction between arguments that show the causes of a thing and arguments that do not. Mancosu shows that a debate originating in the Renaissance about whether mathematical proofs provided true causes for their theorems continued into the seventeenth century and influenced many of the key mathematicians. In particular, he shows that many mathematicians were of the opinion that indirect proofs were inferior to direct proofs because only the latter gave true causes. For example, Descartes in a letter to Marin Mersenne (1588–1648) argued that Fermat's method of maxima and minima was a random procedure "founded on the method of demonstration that leads to the impossible and which is the least esteemed and the least ingenious of all those that one makes use of in mathematics" (Mersenne 1962, 17). The preference for explanatory

A History of Mathematical Impossibility. Jesper Lützen, Oxford University Press.
 DOI: 10.1093/oso/9780192867391.003.0007

proofs was a strong argument for the new infinitesimal methods, because they avoided the double indirect proofs characteristic of the ancient exhaustion proofs. However, it hindered proofs of impossibility because they are usually indirect. For this reason, Antoine Arnauld (1612–94) in the Port-Royal Logic allowed indirect proofs in cases where direct proofs cannot be found, in particular in connection with negative statements:

> Not that these demonstrations are to be altogether rejected, for we may oftentimes employ them to prove negatives.
>
> (Arnauld and Nicole 1662, quoted from Mancosu 1996)

7.2 Descartes's analytic geometry

Throughout the history of mathematics, with the possible exception of ancient Greece, there has always been a close connection between geometry and algebra. In particular during the Arab Middle Ages and in the European Renaissance geometric problems were often solved by algebraic methods, and algebraic rules were proved by geometric arguments. Around 1600 François Viète (1540–1603) developed a systematic methodology, including a new algebraic notation, that made it possible to express general equations and general solution formulas. Following in his footsteps, Pierre de Fermat (1601?–65) and René Descartes (Figure 7.1) suggested how to identify equations in two unknowns with curves whose coordinates satisfy the equations. For example, $y = ax^2$ represents a parabola. This is the central idea of the so-called analytic geometry.

Descartes's account of analytic geometry became particularly influential, and plays a central role in the history of the impossibility of the classical problems. It was published as an appendix entitled *La Géométrie* to one of his major philosophical works, *Discours de la Méthode* (Descartes 1637). The appendices were meant to illustrate how Descartes's philosophical method could be applied, and thus have a strong methodological flavor. In particular in *La Géométrie*, Descartes explained his algebraic method of geometric problem solving.

According to Descartes one solves a geometric problem by assigning letters to the given and the unknown line segments and formulating the relations given in the problem as algebraic equations. Then one combines and simplifies the equations until an equation is found in one of the unknowns only. If this equation is a quadratic equation Descartes showed how the unknown line segment can be constructed by ruler and compass. Thus, it is a plane problem

Figure 7.1 René Descartes (1596–1650)
Reproduced by Jim Høyer DMS

in Pappus' terminology that Descartes adopted. If the equation is of degree higher than 2 it is according to Descartes impossible to solve the problem by plane means. If the final equation is of third or fourth degree, Descartes showed that the unknown can be constructed by the intersection of conic sections, more precisely by the intersection of a circle and a parabola. Following Pappus, he called such problems solid. If the equation is of degree higher than four the problem is, according to Descartes, not solid. If the final equation is of degree five or six, Descartes showed that the unknown can be found by intersecting a circle with a curve of degree 3, the so-called Cartesian parabola. And again he claimed that if the degree is higher than 6 such curves are insufficient to solve the problem. And without going into details he claimed that the process can continue in this way.

Descartes gave complete demonstrations of the results, stating that polynomials of a given degree can be constructed by intersecting the curves mentioned above. His arguments of the corresponding impossibility claims were either lacking or much less convincing. For example, he dealt with the ruler and compass case in the following way:

> And if [the end equation] can be solved by ordinary geometry, that is, by the use of straight lines and circles traced on a plane surface, when the last equation shall have been entirely solved there will remain at most only the square of an unknown quantity, equal to the product of its root by some known quantity, increased or diminished by some other quantity also known. Then this root or unknown line can easily be found. For example, if I have $z^2 = az + b$ ...
>
> (Descartes 1637, 302)

And then he goes on to perform the construction of the solution z of the mentioned equation by a simple ruler and compass construction, followed by the other possible cases of quadratic equations.

In the above quote Descartes starts by formulating the theorem that if the geometric problem is solvable by ruler and compass, then the end equation is a quadratic equation (the impossibility result formulated in contraposed form), but then goes on to prove the converse, namely that if the final equation is quadratic, then the problem can be solved by ruler and compass. Rather than accusing Descartes of making a blatant logical error by confusing a statement and its converse, we may interpret the text as a rhetorical trick aimed at convincing his readers. But the fact remains that Descartes never proved that if the end equation is of degree higher than 2, the problem is impossible with ruler and compass.

Does that matter? Yes it does, because just as Pappus, Descartes insisted that one must always use the simplest means of construction of a problem. For example, if a problem can be constructed using only ruler and compass it is a plane problem, and it would be a methodological mistake to construct it using conic sections or other more complicated curves.

Descartes's account of the true method of construction of geometric problem thus rests on a correlation between the end equation and the geometric means necessary to solve it. But what is this end equation more precisely? Recall that this equation came about by combining and reducing the equations expressing the conditions of the problem. But which means are allowed in these algebraic reductions? It is a thorny question that I shall only discuss briefly.[1]

It is clear that Descartes assumed that his end equation is irreducible in some sense, but the concept of irreducibility was at this time rather fluid. In some arguments it looks as if Descartes irreducibility corresponds to irreducibility over $\mathbb{Q}[X]$. In other arguments Descartes accepts square roots in the coefficients and thus reduces over $K[X]$ where K means the numbers obtained from the rational numbers by adjoining square roots (the numbers constructible by ruler and compass). Of course, that cannot be true in general, because in that case the end equation cannot be a quadratic equation. As it were, Descartes gave many examples of how to reduce equations, but never gave any arguments for the irreducibility of the equations he ended up with.

Connected to this problem is the discrepancy between Descartes's method of solving problems and the classical method of construction. As we

[1] For a more extended but not conclusive discussion see Lützen (2010).

saw earlier, Descartes's method first makes algebraic reductions and then constructs the root of the resulting polynomial equation by one intersection between two algebraic curves. This is very different from a classical construction, which consists of a succession of geometric steps. In each step one draws a number of circles and lines (or other curves in case of a non-plane problem) through (and with centers at) the given points of the problem and the intersection points of the lines and circles from the previous steps. This successive nature of classical ruler and compass constructions is not captured very well by Descartes's method. Indeed, some of the steps in the classical construction may correspond to algebraic steps in Descartes's reduction of the equation while other steps correspond to the final geometric construction of the solutions of the end equation. However, according to Descartes, the nature of the problem only depends on the final equation.

To exemplify this problem let us look at the problem of dividing an angle into nine equal parts. This problem can be formulated by a ninth degree equation, which is irreducible over both the rational and the constructible numbers. Thus, according to Descartes it must be constructed by the intersection of a circle and a fifth degree curve. Yet, it is clear that the problem can be solved by first trisecting the given angle and then trisecting the resulting angle. As also proved by Descartes (see Section 7.3) the two trisections can be done by intersecting conic sections, so that the problem according to Pappus is a solid one, at variance with Descartes's method of solution. This illustrates how Descartes's method does not reflect the usual successive character of geometric constructions.

Descartes's general claims concerning the impossibility of problems with plane, solid, or other means are thus highly problematic. As we shall see in Section 7.3, there are reasons to believe that Descartes, in fact, realized at least some of the methodological problems.

7.3 Descartes on the duplication of a cube and the trisection of an angle

In *La Géométrie* Descartes deduced the equations corresponding to the trisection of an angle and the duplication of a cube, or more generally the construction of the two mean proportionals. Assume a and b are two given line segments and z and y are their two mean proportionals, i.e.,

$$\frac{a}{z} = \frac{z}{y} = \frac{y}{b}. \qquad (7.1).$$

As shown in Chapter 4 a short deduction gives that z is the solution of the simple cubic equation

$$z^3 = a^2 b\,. \qquad (7.2).$$

A slightly more complicated argument led Descartes to the equation

$$z^3 = 3z - q \qquad (7.3)$$

for the chord z of an arc in a unit circle which is one-third of the arc having a chord q.[2] From his general method for constructing a solution of a cubic equation, Descartes then showed how the two problems can be solved by a solid construction intersecting a circle and a parabola.

In order to show that the two problems are indeed solid and not plane, one would have expected that Descartes would have given the following simple argument based on his general remarks on problems and their constructions: Since the two problems lead to cubic equations rather than quadratic equations, they cannot be solved by plane means. However, Descartes never explicitly mentioned this simple argument. One could argue that it is because it is so obvious that he did not find it worth mentioning. Yet, toward the end of the book, he admitted that he did not pretend he had given the reasons for his impossibility statements:

> It is true that I have not yet stated my grounds for daring to declare a thing possible or impossible, but if it is remembered that in the method I use, all problems which present themselves to geometers reduce to a single type, namely to the question of finding the values of the roots of an equation, it will be clear that a list can be made of all the ways of finding the roots, and that it will then be easy to prove our method the simplest and most general.
>
> (Descartes 1637, 401)

So concerning his claim that his method will yield the simplest solution of geometrical problems he admitted that he had not given sufficient arguments yet. His suggestion that it can be shown by inspecting all possible ways of solving the end equation must have appeared weak even to Descartes himself, in particular since his claim that one can make a list of all possibilities seems questionable if not outright wrong. However, Descartes did not leave the problem

[2] It corresponds to the cubic equation found two centuries earlier by al-Kāshī (see Section 6.3).

there. On the subsequent half-page, he finally produced an argument that was intended to show that his solutions gave the simplest construction of geometric problems. I shall quote it in full, because it very explicitly presents an argument for the impossibility of constructing the two classical problems by ruler and compass.

> Solid problems in particular cannot, as I have already said, be constructed without the use of a curve more complicated than the circle. This follows at once from the fact that they all reduce to two constructions, namely, to one in which two mean proportionals are to be found between two given lines, and one in which two points are found which divide a given arc into three equal parts. Inasmuch as the curvature of a circle depends only upon a simple relation of all its parts to the point that is its center, the circle can only be used to determine a single point between two extremes, as, for example, to find one mean proportional between two given lines or to bisect a given arc; while, on the other hand, since the curvature of the conic sections always depends upon two different things, it can be used to determine two different points.
>
> For a similar reason, it is impossible that any problem of degree more complex than the solid, involving the finding of four mean proportionals, or the division of an angle into five equal parts, can be constructed by the use of one of the conic sections. I therefore believe that I shall have accomplished all that is possible when I have given a general rule for constructing problems by means of the curve described by the intersection of a parabola [a Cartesian parabola] and a straight line, as previously explained.
>
> (Descartes, 1637, 401–2)

This argument or set of arguments call for comments. First, it is worth pointing out that Descartes did not talk about *proving* the impossibility statements but about giving the reasons for them. This may reflect his adherence to the idea that mathematics must give the causes for its theorems and it would explain why he did not attempt an indirect proof. If the above argument was not meant as a rigorous proof but as an uncovering of the causes, we cannot expect a rigorous line of thought. Still I think that even as a non-rigorous argument, it is difficult to follow.

Now let us turn to the argument intended to show that angle trisection and two mean proportionals cannot be constructed by ruler and compass. Although Descartes had no well-defined concept of curvature, it seems reasonable to say that the curvature of the circle depends on one thing (the radius), whereas the curvature along a conic section depends on two things, e.g., the distance of the point from the two foci. However, it is hard to comprehend what this curvature has to do with the number of points that the curves can be used to determine. Descartes's claim that a circle can only be used to determine a single point is particularly remarkable. First, he knew that in general a line and a circle or two circles intersect in two points if at all. Second, he knew that one can solve the trisection of a line segment by ruler and compass. Third, it is not clear why Descartes seems to think that the construction should lead to both of the mean proportionals (z and y in (7.1)) or to both of the two trisection points of the angle. Indeed, his own construction of the chord of the third of an angle with a given chord does not provide the chord of two-thirds of the angle. Instead, Descartes's construction determines the three roots of Eq. (7.3). As he pointed out himself (Descartes 1637, 397) the three roots represent the chord of a third of the given arc, the chord of a third of the remaining part of the whole circle, and minus the sum of those two chords.

Let me now turn to the argument that solid problems in general cannot be solved by ruler and compass. First, it is clear that Descartes at this point took "solid problems" to mean problems that lead to an end equation that is of degree 3 or 4. As mentioned in the extract, Descartes had earlier shown how one can reduce the solution of the general cubic equation to the problem of constructing two mean proportionals, corresponding to the extraction of the third root in Cardano's formula (see Section 10.3). This method does not work in the so-called irreducible case (see Section 10.3). However, in this case Descartes, following Viète, showed how the solution could be obtained from an angle trisection. Moreover, following Ferrari, Descartes had also reduced the solution of the fourth degree equation to a cubic equation, the so-called resolvent. In the extract under scrutiny he then used these reductions to argue for the impossibility of solving the general equation of degree 3 and 4 by ruler and compass. The argument seems to go as follows: If the problems of angle trisection and of constructing the two mean proportionals were possible by ruler and compass, the solution of the general third and fourth degree equation would be possible by these means because the constructions can be reduced to those two specific cases. However, by the argument discussed above, the angle trisection and two mean proportionals cannot be found by ruler and compass, and therefore the general third and fourth degree equation cannot be solved by these means either. Of course, this argument is invalid, because it confuses a statement with its inverse. However, it is not obvious how Descartes's argument can be interpreted differently. Indeed, just because the general problem can be reduced to special cases that cannot be constructed by ruler and compass it is not clear why cubic equations other than those two might not be solvable by these means. And it is quite clear that Descartes wanted to say more than the general cubic is not solvable by ruler and compass because there are two cases that cannot be solved that way. If that had been all he wanted to say, he would not have needed to invoke the reduction of the general cubic to the two special cases.

The last part of the quote, where Descartes generalized the impossibility argument to equations of degrees higher than 4, are equally problematic, but I shall not dwell on it (see Lützen 2010, 24).

It is remarkable how Descartes's impossibility argument differs in its structure from the modern one. We show by algebraic arguments that one cannot solve an irreducible cubic (or more generally, an irreducible equation of a degree which is not a power of 2) by square roots, and since the problem of the two mean proportionals and the trisection of an angle corresponds to irreducible cubic equations we conclude that these two classical problems are impossible with ruler and compass. Descartes argued conversely that these two classical problems are impossible with ruler and compass and since every cubic equation can be reduced to these two problems, general cubics are not solvable by ruler and compass.

What is even more remarkable is that in this final argument only geometry is involved. All Descartes's constructive arguments concerning the solution of geometric problems rely heavily on the corresponding equations. Here in the final impossibility argument the equations play no role. In fact, if he had involved the equations into the above arguments, they would have appeared even weirder. Indeed, these equations do not have two but three roots, so in principle a construction of the roots should give three values, and not two. And the three solutions of the cubic equations do not even include the other point mentioned in Descartes's geometric argument. For example, Eq. (7.3) has one real root, namely the first of the two mean proportionals z in (7.2), and two complex roots. The second mean proportional y, to which Descartes's geometric impossibility proof appeals, is not a root of Eq. (7.3) but of a different equation, namely

$$y^3 = ab^2.$$

As mentioned earlier, a similar remark applies to the equation for angle trisection. Therefore, if Descartes had invoked the equations corresponding to the two problems that had originally led him to the construction of the solution, it would have been very strange why he thought that the construction should yield the other mean proportional and the second division point of the angle rather than the two other roots of the equation.

For Descartes's successors, the equations became important also for the proofs of impossibility.

7.4 Descartes's contributions

Although Descartes did not succeeded in proving the impossibility of the duplication of a cube or the trisection of an angle in a way that was generally accepted by his successors, he made important and lasting contributions to the problem.

1. He considered the impossibility problems to be mathematical problems that needed a proof. We saw that Pappus and other Greek mathematicians had believed that the problems were in principle impossible with ruler and compass, but they had dealt with the impossibility as a metamathematical question rather than a mathematical question that called for a proof.
2. He translated the problems into algebraic equations.
3. He generalized the impossibility theorems to a whole class of similar problems, namely those problems whose end equation is of degree 3 or higher.
4. He showed that quadratic equations could be solved with ruler and compass and claimed that conversely, no equations of higher degree could be solved with these means. Though the last claim turned out to be wrong, the close relationship between quadratic equations and solvability with ruler and compass remained important.
5. He emphasized the importance of some kind of irreducibility of the polynomial for the question of its constructibility.

However, his way of dealing with irreducibility was insufficient for the purpose of establishing an algebraic impossibility proof, and he lacked an algebraic translation of the successive process of construction by ruler and compass. In the end, he seems to have admitted that his algebraic approach to the impossibility proof was not conclusive and so he fell back on a puzzling purely geometric argument.

7.5 The eighteenth century

In 1775 the French Academy of Sciences declared that they would no longer evaluate circle quadratures, cube duplications, angle trisections, and perpetual motion machines. By way of motivation the Secretary of the Academy M. J. A. Nicolas de Condorcet (Figure 7.2) wrote about the duplication of a cube, or more generally the two mean proportionals:

> One sees then [by applying algebra to geometry] that the problem reduces to the solution of an equation of the third degree which has two imaginary roots; that one can only construct it by the intersection of a straight line and a curve of the third degree or by the intersection of two curves of the second degree ... ; one knows the simplest methods and one sees that it is useless to try to solve it using only the circle and the straight line.
>
> (Condorcet 1775/8, 61–2)

He dealt with the angle trisection in a similar manner concluding:

> For a long time the analysis [the modern geometers] have given of this problem is complete and leaves nothing to be desired.
>
> (Condorcet 1775/8, 62)

Figure 7.2 M. J. A. Nicolas de Condorcet (1743–94)
Reproduced by Jim Høyer DMS

Thus, the highest scientific authority maintained that the impossibility of the two problems with ruler and compass was a well-proven fact. The proof sketched in Condorcet's motivation went back to Jean Étienne Montucla (1725–99), who discussed the two problems in his *Histoire des recherches sur la quadrature du cercle* (1754) and in his famous two-volume *Histoire des Mathématiques* (1758, 2nd edn 1799). In both works, Montucla recorded many of the ancient as well as more recent constructions by Descartes, Isaac Newton (1643–1727), and René-François de Sluse (1622–85) of the two problems while emphasizing the impossibility of solving them with ruler and compass alone. About the duplication of a cube he stated that he would treat

> the solutions such as the nature of the problems admit, that is solutions in which one uses some other curve than the straight line and the circle or some instrument other than ruler and compass. *For one proves today*, and the ancients were not ignorant about it, that one cannot solve it with the aid of the ordinary geometry.
>
> (Montucla 1758, vol 1, p. 188, my italics)

In a similar vein, he declared about the trisection of an angle or the division of an angle in a given ratio:

> It was in vain that one tried to solve one of these problems by plane geometry. Of the same nature as the duplication of the cube, they require the help of a higher kind of geometry or the use of some instruments other than the ruler and the compass.
>
> (Montucla 1758, vol 1, p. 193)

Both Montucla and Condorcet emphasized that it was the recent development of analytic methods that made the impossibility proof possible. Thus, Condorcet wrote about the two mean proportionals:

> But one did not have, and one could not have had, a complete analysis of the problem until one had found the principles of the application of algebra to geometry.
>
> (Condorcet 1775, 61)

In a similar vein, Montucla stated:

> This impossibility is founded on the theory of equations and the nature of geometric curves.
>
> (Montucla 1754, 274–5)[3]

[3] The claim that an impossibility proof of the classical problems would be impossible without algebraic techniques is often repeated, e.g., by Coolidge (1940, 53). It is a typical meta statement that no one seems to find worthy of a proof.

Developing this idea, Montucla explained in general terms the nature of the recent proofs:

> [the two classical problems] owe their complete solution to the modern geometry. In fact only the light provided by this geometry has enabled us to see that they are of a nature that cannot generally be solved by plane geometry; This was a necessary point to demonstrate before breaking off their [the trisectors, etc.] efforts to solve them in this way. But the modern analysis lifts all doubt in this respect. Moreover, what the ancients have done concerning this subject compared with the inventions of the geometers of the last century is as a feeble day next to a great light.
>
> (Montucla 1754, 273–4)

The very explicit enlightenment rhetoric visible in this quote can be found in many other eighteenth-century works on the classical problems. In view of Montucla's repeated claims that the moderns had now proved the impossibility of the two problems, it is conspicuous that he did not refer to any concrete proof of impossibility, nor did he mention any mathematician who had given such a proof. Instead, he gave a proof of his own. His proof of the "impossibility of constructing these problems in general without using curves that are more composed [complicated] than the circle" (Montucla 1754, 274) was Cartesian in the sense that it was "based on the theory of equations and the nature of geometric curves" (Montucla 1754, 274–5). In particular, it was based on two principles:

1. That every polynomial equation has as many roots as the degree of the polynomial (the fundamental theorem of algebra).
2. "The second principle is that an equation can only be constructed geometrically, that is by a certain procedure that is not the result of groping, by the aid of two curves that can intersect each other in as many points as the degree of the equation contains units." (Montucla 1754, 275)

This second principle was the fundamental guideline in a research program that Bos has called "the construction of equations" (see (Bos 1984). As pointed out by Bos, this research program began with the Latin translation of Descartes's *La Géometrie* and flourished for the next century. And although Montucla wrote his books around the time when Bos has noted a decline in the research tradition, Montucla took its central principle for granted. The only point he tried to argue for was the claim that the construction must furnish not just one of the roots of the equation but must furnish all of them, i.e. that the two curves used in the construction must intersect each other at a number of points equal to the degree of the equation:

> because this construction does not concern one [root] rather than another one, since the givens are the same for both of them and these [the givens] are the only ones that can modify [the construction].
>
> (Montucla 1754, 276)

The idea of this argument seems to be that the curves used in the construction must be formed from the givens of the problem (the coefficients of the equation) and since these are the same for two different roots, the curves used to construct each of the roots must be the same.

According to Montucla, this argument establishes that the number of intersections of the two curves used in the construction of the solution must be equal to the number of roots in the equation, i.e., to its degree. Montucla argued that this must hold true even if some of the roots of the equation are imaginary, because imaginary roots are just "intersections that particular limitations have rendered impossible" (Montucla 1754, 277).

Having drawn attention to the two principles mentioned above, Montucla went on to formulate the equations corresponding to the two classical problems. In contrast to Descartes, he explicitly commented on their irreducibility. As far as the problem of the two mean proportionals was concerned, he simply claimed without proof that its equation

$$z^3 = a^2b$$

is irreducible unless b is a cube times a. Concerning the trisection of an angle he argued in a different way declaring, that its equation must be of the third degree except for particular values of the angle. In a rather long passage, he argued that the three different roots of the trisection-equation (7.3) that Descartes had already characterized geometrically must necessarily all be solutions of this equation, and thus it must have at least three roots.

Having established that the two problems lead to cubic equations, he could then apply his two principles:

> it is evident that one cannot construct them by employing only curves capable of giving less than three intersection points.
>
> (Montucla 1754, 283)

Thus using (but not explicitly mentioning) the obvious fact that two lines or two circles or a line and a circle can intersect in at most two points, he concluded:

> Those who try to combine circles and straight lines in order to find this solution fruitlessly waste their time and their night's sleep.
>
> (Montucla 1754, 283–4)

In this way, Montucla had used the method of construction of equations to put forward a proof of the impossibility of solving the two classical problems by ruler and compass, and he had enlightened both professional mathematicians and amateurs about this impossibility.

Condorcet apparently endorsed Montucla's proof. Indeed Condorcet's impossibility argument mentioned earlier was clearly a sketchy version of Montucla's. Other mathematicians of the eighteenth century may not have been as easily convinced. For example, in his article "Trisection" in the famous *Encyclopédie*, the leading French mathematician Jean-Baptiste le Rond d'Alembert (1717–83) did not give an impossibility proof of a solution with ruler and compass neither of this problem nor of the duplication of a cube (d'Alembert 1751–66, Trisection). He did not even refer to Montucla's proofs, although he praised Montucla's book on the history of the quadrature of a circle in his article "Quadrature du cercle" in the *Encyclopédie* (d'Alembert 1765, Quadrature du cercle). He may have questioned the correctness of Montucla's proof. Indeed, he wrote about the problem of trisecting an angle by ruler and compass that "this is one of those problems that one has tried in vain to solve for two thousand years, and which in this respect, together with the duplication of the cube, can be compared to the quadrature of the circle." As we shall see, he did not believe that the quadrature of a circle had been proved impossible so his comparison of the three problems indicates that he did not think the impossibility of the angle trisection and cube doubling had been proved either. Still, according to Jacob (2005), he was one of the most outspoken supporters of the ban of these questions from the academy.

7.6 Montucla and Condorcet compared with Descartes

Montucla's impossibility argument and Condorcet's survey of it are much clearer than Descartes's arguments. First, in contrast to Descartes, who discussed the impossibility of the problems in many places of *La Géométrie*, Montucla devoted a separate section to the explicit formulation and proof of the impossibility. Secondly, in contrast to Descartes's final discussion of the "reasons" for the impossibility, Montucla's proof was based on the equation of the problem and on an algebraic formulation of the method of solution, namely the intersection of two algebraic curves. In this way Montucla's proof was thoroughly algebraic in nature whereas Descartes's proof was geometric.

In the hands of Montucla and Condorcet, Descartes's final argument was transformed dramatically. Instead of arguing that the construction of the mean proportionals and the trisection points should necessarily yield two points, as Descartes had done, the introduction of the equation made Montucla and Condorcet argue that the construction should yield three values, namely the three roots of the cubic equation. And while Descartes used a strange curvature argument to "show" that a circle can only determine one point, Montucla and Condorcet could appeal to the more obvious fact that circles and straight lines have at most two intersection points. Thus, by changing Descartes's proof into an algebraic one and raising all the entering numbers by one, Montucla (and Condorcet) succeeded in clarifying the argument. One can say that their resulting proof was more Cartesian than that of Descartes himself.

By clarifying the impossibility proof, Montucla also exposed its shortcomings more clearly than Descartes had done. Indeed, to a modern reader, who is not as strongly influenced (brainwashed) by the Cartesian "construction of equations" research program, it is quite obvious that the fundamental basis for Montucla's impossibility proof is completely misconstrued. It is, of course, entirely possible to declare, as Montucla and Condorcet did, that the only allowable construction of a root of an equation is an intersection between two algebraic curves. However, it is also clear (at least to a modern reader) that this concept of correct problem solving is at odds with the classical Euclidean method of construction. In the classical tradition, one does not limit oneself to one intersection between two curves. One is allowed to make a long succession of intersections, leading to the desired point or line segment. By implying that their method of solution somehow included the classical ruler and compass solutions as a special case, Montucla and Condorcet seem to have convinced many of their contemporaries that the impossibility was a proven fact.

However, to a modern reader it is quite clear that the only thing Montucla and Condorcet really argued was that one cannot solve the classical problems by making one intersection between a circle and a straight line or between two circles or two straight lines. However, that would hardly surprise a serious angle trisector or discourage him from trying to find a more involved ruler and compass construction.

8
Circle Quadrature in the Seventeenth Century

8.1 "Solutions" and positive results

In the Renaissance and the early modern period, the quadrature of a circle was by far the most studied of the three classical problems. Several well-known mathematicians and philosophers believed they had solved the problem, including Nicholas of Cusa (1401–64), Oronce Fine (1494–1555), Christopher Clavius (1538–1612), Joseph Justus Scaliger (1540–1609), Christian Longomontanus (1562–1647), Gregoire de Saint-Vincent (1584–1667) (Figure 8.1), and Thomas Hobbes (1588–1679). Some of the constructions used the quadratrix and tried to argue that this curve was acceptable; other constructions turned out to be only approximations. But none of the solutions were widely accepted and several of them led to public debates. For example, Scaliger's circle quadrature was refuted by Ludolph van Ceulen (1540–1610) (Hogendijk 2010), Longomontanus' was refuted by John Pell (1611–85) and others (van Maanen 1986), and Hobbes' led to a long controversy with Wallis (Jesseph 1999). In 1634 the unofficial coordinator of the French mathematical community Marin Mersenne summed up the situation regarding the question "the quadrature of the circle, is it impossible?" in the following words: "This problem is extremely difficult, for one can find excellent geometers who claim that it is not possible to find a square whose surface is equal to that of the circle, and others who claim the opposite" (quoted from Mancosu 1996, 79).

This opinion changed slowly over the next one and a half centuries. By 1780 the official opinion expressed by the French Academy of Sciences was that the problem was impossible, but there was still no general agreement as to whether this impossibility had been proved. In this chapter, we shall consider the research that led to this change of opinion.

However, before we turn to the impossibility question, let us briefly have a look at the many positive discoveries made concerning the circle quadrature. From Archimedes' time the value of π was evaluated with ever greater accuracy by Greek, Chinese, Indian, and Arab scholars. We have already pointed

A History of Mathematical Impossibility. Jesper Lützen, Oxford University Press.
 DOI: 10.1093/oso/9780192867391.003.0008

Figure 8.1 Frontispiece to Grégoire de Saint-Vincent's *Opus geometricum quadraturae circuli et sectionum coni decem libris comprehensum* (1647). In the background to the right angels are engaged in squaring the circle, or circulating the square. At the bottom left the mathematician points to Archimedes' theorem stating that a circle is just as large as a right-angled triangle with the two smallest sides equal to the radius and the circumference of the circle

Reproduced by Jim Høyer DMS

out that in 1424 Jamshīd al-Kāshī (1380–1429) determined π to 16 decimal places. After that, the Europeans took over and around 1600 Ludolph van Ceulen computed the first 35 decimal places. All these calculations were done using the technique that Archimedes had used. After van Ceulen, higher accuracy was obtained using more refined numerical methods and new infinite analytical expressions for π. The first improvement of the numerical methods was made by Christiaan Huygens (1629–95) in 1654 in his *De Circuli Magnitudine Inventa.* He proved various theorems that made it possible to tease out more decimals of π from the values of an inscribed and a circumscribed regular polygon with a given number of sides.

The first infinite analytic expression for π was discovered by Viète (1593). It was an infinite product, involving square roots. It is just Archimedes' method written out analytically. A simpler infinite product was discovered by John Wallis (1616–1703) in his main work *Arithmetica Infinitorum* (1656),

$$\square = \frac{4}{\pi} = \frac{3{\cdot}3{\cdot}5{\cdot}5{\cdot}7{\cdot}7\cdots}{2{\cdot}4{\cdot}4{\cdot}6{\cdot}6{\cdot}8\cdots}, \tag{8.1}$$

which is Wallis' notation for the ratio of the square of the diameter in a circle to the circle itself. The meaning of this formula is that if one takes the same number of factors in the numerator and the denominator, the resulting fractions will converge to $4/\pi$ when the number of factors goes to infinity.

The first infinite series for π was Leibniz's series

$$\frac{\pi}{4} = \frac{1}{1} - \frac{1}{3} + \frac{1}{5} - \frac{1}{7} + \cdots, \tag{8.2}$$

which Gottfried Wilhelm Leibniz (1646–1716) himself called "the arithmetical circle quadrature." This series had already been discovered around 1400 by the Indian mathematician Madhava, who worked in in the Kerala school, but that was not known in Europe in the seventeenth century.

8.2 Descartes on the quadrature of a circle

The duplication of a cube and the trisection of an angle fitted very well into Descartes' method of solving geometric problems. They could be translated into polynomial equations that could be solved by the intersection of conic sections. The quadrature of a circle did not fit into this scheme so Descartes did not discuss this third classical problem in *La Géométrie.* Still, Bos (2001) and Mancosu (1996) have argued that the problem had a decisive influence on Descartes' method for geometric problem solving. Indeed, his view of the problem of squaring or rectifying a circle was instrumental in his rejection from geometry of what he called mechanical curves, i.e., curves that are not algebraic. Following Aristotle, Descartes denied that curved lines could be compared exactly with straight lines (Mancosu 1996, 77). For this reason, curves like the quadratrix and the spiral cannot be admitted as exact curves because their construction involves the coordination of a rectilinear motion with a circular motion, which is in Descartes' opinion necessarily inexact. For this reason, he rejected the known constructions of the quadrature of a circle. This line of argument has been reconstructed by Bos (2001, 342). Mancosu has found a slight variation of this argument in a letter from Descartes to Mersenne from 1629 (Mancosu 1996, 74ff.): If one accepts the quadratrix and the spiral as exact geometric curves, one can use them to find a rectification of the circle, but that is known to be impossible. Thus the curves are not acceptable.

A third argument against the quadratrix was put forward by Descartes in connection with a rejection of Clavius' circle quadrature. Clavius had argued that the quadratrix was a geometric curve on par with the conic sections because one can construct an infinity of points (in fact, a dense set of points) on it with geometric means, i.e., with ruler and compass. And since the quadrature of a circle could be solved using the quadratrix he had declared this problem to be solved by geometric means. Descartes disagreed. He pointed out that there was a difference between the conic sections and the quadratrix: On the conic sections one can construct every point, but on the quadratrix one can only construct selected points. For example, one cannot construct the points corresponding to the trisection of an angle or the point that leads to the quadrature of a circle.

In a letter of 1638 to Mersenne, Descartes even suggested that the problem was inappropriate:

> For, in the first place, it is against the geometers' style to put forward problems that they cannot solve themselves. Moreover, some problems are impossible, like the quadrature of the circle.
>
> (quoted from Mancosu 1996, 78)

La Gémoétrie had a great influence on the subsequent development of geometry, but Descartes' ban on mechanical curves and his assertion that the straight and the curved were incompatible soon lost their credibility. First, around 1657 William Neile (1637–70) and Hendrick van Heuraet (1634–60?) discovered that it is possible to determine the arc length of the so-called semi-cubical parabola[1] algebraically. The following year Christopher Wren (1632–1723) proved that the arc of a cycloid is equal to four times the diameter of the generating circle. These discoveries knocked holes in Aristotle's and Descartes' claim that one could not rectify curved lines. Moreover, Leibniz soon extended Descartes' analytic methods to include mechanical curves, or transcendental curves as he called them (Blåsjö 2017).

8.3 Wallis on the impossibility of an analytic quadrature of a circle

Most seventeenth-century impossibility arguments concerning the quadrature of a circle were comments to positive results. This holds true of the argument presented by John Wallis (Figure 8.2) in his *Arithmetica Infinitorum* (1656) intended to show that his infinite product for □ or π was the best possible solution of the quadrature of a circle; best possible in the sense that it is impossible to write □ or π as a finite expression using only rational operations and root expressions. In Chapter 12 we shall see how Wallis argued that impossible problems had earlier led to a successive extension of the number system by inventing new notation. The aim of this digression was to prepare the reader for the next extension, namely his invention of the notation □ which allows us to consider this ratio as a true number.[2] In order to argue that this was really an extension of the number concept, he needed to argue □ that could not be expressed using known notation (i.e., $+$, $-$, $\cdot$, $:$, $\sqrt[n]{\ }$).

His argument was not based on his infinite product (8.1) but on the way he had found it: In a process of bold "induction" and interpolation Wallis had filled out a doubly infinite table containing the values of the ratio of a unit square and the area under the curves

$$y = \left(1 - x^{1/p}\right)^q \qquad (8.3)$$

in the interval (0,1) for values of p and q equal to -½, ½, 1, 1½, 2, 2½, 3, ... The entry of this table corresponding to $p = q = ½$ is precisely the ratio □ that he wanted to determine. He argued that the entries of the table having an integer value of either p or q (called the even sequences in the quote below) would be rational fractions, whereas the entries having both p and q half-valued would be a rational fraction times □.

[1] The curve with the equation $ay^2 = x^3$.

[2] Wallis also used other symbols for the new number.

Figure 8.2 John Wallis (1616–1703)
Reproduced by Jim Høyer DMS

Wallis' impossibility argument was based on his expression of the areas in question as ratios of sums. Indeed, if one divides the abscissa interval (0,1) into n equal parts and sums the ordinates y corresponding to the ordinates of the division points, the sum will, according to Wallis' method of indivisibles, represent the area under the curve expressed by (8.3). The sum of an equal number of indivisibles equal to the longest ordinate (i.e., 1) will likewise represent the area of the unit square. The ratio of these two sums will thus represent the inverse of the ratio □. It can be written in modern notation:

$$\frac{\sum_{i=0}^{\infty}\left(1-\left(\frac{i}{n}\right)^{1/q}\right)^{p}}{n+1}=\frac{\sum_{i=1}^{\infty}\left(n^{1/q}-i^{1/q}\right)^{p}}{(n+1)\cdot n^{p/q}}. \tag{8.4}$$

In order that this ratio shall represent the exact ratio of the areas, n must be taken to be infinite.

With these preliminaries in place, we can begin to understand Wallis's argument:

> And indeed I am inclined to believe (what from the beginning I suspected) that this ratio we seek [□] is such that it cannot be forced out in numbers according to any method of notation so far accepted, not even by surds [...] so

that it seems necessary to introduce another method of explaining a ratio of this kind, than by true numbers or even by accepted means of surds.

And indeed this, whether opinion or conjecture, seems to be confirmed here, since if we have the appropriate formula, for any even sequence (in the table of proposition 184) [the above-mentioned table] so also we might have obtained a formula of this kind for any odd sequence; then just as for the formulae for the even sequences we have taught how to investigate the ratio of finite series of first powers, second powers, third powers, fourth powers, etc, to a series of the same number of terms equal to the greatest of those (in the comment to proposition 182) so by formulas of the same kind for odd sequences, it would seem there could be investigated similarly the ratio of finite series of second roots, third roots, etc. to a series of the same number of terms equal to the greatest of these: why this is not to be hoped for, moreover, we showed in the comment to Proposition 165.

(Wallis 1656, 161)

In the last-mentioned comment Wallis had discussed sums similar to (8.4) where $p = q = ½$. He had pointed out that the numerator was a sum of $n + 1$ different square roots, whereas the numerator is an integer. He explicitly calculated the ratio (8.4) for $n = 6, 10, 12$, and 20. For example, for $n = 10$ he got

$$\frac{\sqrt{100-0}+\sqrt{100-1}+\sqrt{100-4}+\sqrt{100-9}+\cdots+\sqrt{100-81}+\sqrt{100-100}}{110}, \tag{8.5}$$

which according to Wallis (p. 123) "cannot be written otherwise more briefly than"

$$\frac{24+3\sqrt{11}+4\sqrt{6}+\sqrt{91}+2\sqrt{21}+5\sqrt{3}+\sqrt{51}+\sqrt{19}}{110}. \tag{8.6}$$

And in the same way, as more parts of the radius are taken, so the expression for the ratio necessarily becomes more intricate; and indeed requires repetition of almost all the roots, since they happen little, indeed rarely, and only as if by chance, to be commensurable either with rational numbers or with each other.

(Wallis 1656, 123)

So when n grows the expression will contain more and more square roots and thus seems to end up being less and less expressible.

> Therefore if the radius [...] is taken in infinitely many parts (which it seems must be done for our purposes) the ratio of all sines, to the radius taken the same number of times, that is the quadrant [...] circle to the circumscribed square [...], seems wholly inexpressible.
>
> (Wallis 1656, 124)

What Wallis presents in the above-quoted comment to proposition 190 seems to be conceived as a strengthening of this comment to Proposition 165. Wallis seems to argue that if □ were expressible in terms of a finite number of radicals or surds (i.e., nth root signs), then the expression (8.4) would be thus expressible (for $n = \infty$) not only for $p = q = ½$ but for all integer or "half" values of p and q. Thus, an infinity of different sums of radicals (not just square roots but all other kinds of mth roots) would be reducible to the same finite number of radicals which is "not to be hoped for."

Wallis's argument seems weak in particular in the light of an earlier passage of the *Arithmetica Infinitorum* (a comment to proposition 165), where Wallis had given an example of a similar situation where an infinity of radicals do, in fact, add up to a rational number. While studying the quadrature of curves having an equation

$$y = \sqrt[q]{x},$$

Wallis had expressed the area under the curve in the interval (0,1) as the ratio

$$\frac{\sum_{i=0}^{n} \sqrt[q]{i}}{(n+1)\sqrt[q]{n}}.$$

However, Wallis had argued that for $n = \infty$ this ratio is equal to $\frac{q}{q+1}$, "the infiniteness itself indeed (which seems amazing) destroying the irrationality" (Wallis 1656, 125). At this point of the book, he saw this as a sign that the quadrature of a circle was not out of sight:

> so clearly not all hope was lacking of eventually finding the ratio of a series of universal roots (of augmented or reduced series) to a series of equals. And indeed if not in every case, at least for those so far set out; and perhaps even in those that touch on the quadrature of the circle itself or the ellipse, or also the hyperbola, something may be gained.
>
> (Wallis 1656, 125)

Moreover, something similar happens for all the values in the infinite table corresponding to integer values of q (the even rows in the table). And even when q is half-valued (½, 1½, 2½, ...) the ratio is also rational when p is an integer.

This may be the reason why Wallis in *Arithmetica Infinitorum* did not explicitly claim that he had proved that □ cannot be expressed in terms of finitely many radicals, but only wrote that a finite expression would be too much to hope for. He cautiously referred to the non-algebraic nature of □ as an "opinion or conjecture" that "seems to be confirmed here." Nevertheless, in his later critique of Gregory's argument for this claim, he called his own argument a demonstration:

> as having many years since demonstrated the same myself, though [Gregory] take no notice of it, in my Arithmetica Infinitorum, proposition 190 with ye Scholium annexed.
>
> (Wallis 1668, 285)

Before we turn to Gregory's argument and the subsequent controversy with Wallis, some conceptual clarifications are needed.

8.4 Different quadratures of a circle

The classical answer to a geometric problem was a geometric construction. The question was whether one could find a solution by ruler and compass or conic sections, or whether one needed more complicated curves as the quadratrix. With the introduction of Cartesian analytic geometry the geometric formulation was gradually transformed into an analytic one. In the case of the quadrature of a circle the question became the following: can the area of the circle (or its circumference) be determined from its radius

a. by rational operations (the question of commensurability)?
b. by algebraic operations, either an explicit expression in terms of radicals or implicitly through an algebraic equation?
c. by an infinite number of rational or algebraic operations?

The classical geometric constructions were hardly discussed in the seventeenth century, probably because it was believed that the answer to the analytic problem would somehow reveal the answer to the geometric problem. As mentioned in the introduction to this chapter the positive answers to c. were the main positive result of the seventeenth century concerning the quadrature of a circle. It led to Viète's and Wallis's products, Leibniz's series, and several other infinite analytic expressions of π. Various arguments for the negative answer to b. were the main contribution of the seventeenth century to the impossibility question.

It is necessary to introduce one more distinction, namely between the definite and the indefinite circle quadratures.

1. The definite quadrature deals with the area or circumference of the entire circle.
2. The indefinite quadrature deals with the area or the arc length of a sector $S = OACBO$ of the circle expressed in terms of the radius r of the circle and the chord $x = AB$ of the sector (Figure 8.3)

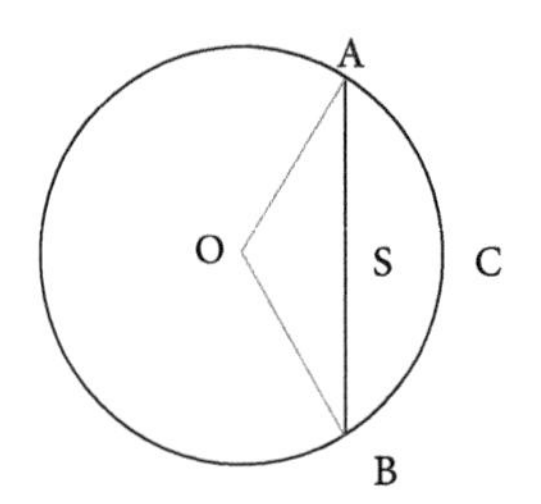

Figure 8.3 A sector of a circle

Wallis dealt with question 1, but most other seventeenth-century mathematicians dealt primarily with the indefinite quadrature 2. The connection between the definite and the indefinite circle quadratures was not clear in the seventeenth century and it gave rise to a great deal of confusion. The reason is that a positive solution to the problem of the indefinite quadrature of a circle can mean three different things:

2a. There exists one algebraic formula in r and x (or one algebraic equation in r, x, and S) that expresses the area S of the sector as a function of its chord x. This question corresponds to the question "Is arcsin an algebraic function (suitably interpreted)?"

2b. For each sector of the circle, there exists an analytic formula in r and x (or one algebraic equation in r, x, and S) that expresses the area S of the sector as a function of its chord x. Here we allow different formulas for different sectors.

2c. There exist sectors S (different form 0) of the circle whose area can be expressed by an algebraic formula in r and x (or an algebraic equation in r, x, and S).

We now know that 2a and 2b are false and 2c is true.[3] Corresponding to these positive answers to the circle quadrature there correspond the following converse impossibility statements:

¬2a. There is no algebraic function (explicit or implicit) expressing the area of a sector S as a function of r and x (arcsin is not an algebraic function).
¬2b. There exists a sector whose area cannot be expressed by an algebraic function (explicit or implicit) of r and x.
¬2c. No sector has an area that can be expressed by an algebraic function (explicit or implicit) of r and x.

It follows from the previous footnote that $\neg 2a$ and $\neg 2b$ are true and $\neg 2c$ is false.
The following implications are obvious for a modern reader:

$$2a \Rightarrow 2b \Rightarrow 2c, \quad \neg 2c \Rightarrow \neg 2b \Rightarrow \neg 2a.$$

Moreover if 1b denotes the existence of an algebraic definite circle quadrature (i.e., an algebraic quadrature of an entire circle) the following implications hold:

$$1b \Rightarrow 2c, \quad \neg 2c \Rightarrow \neg 1b.$$

No other implications connect the definite and the indefinite circle quadratures. I have made these distinctions clear because we will see that they were not always clear to the seventeenth-century actors. In particular, their slippery use of the quantifiers "for all" and "there exist" and their order often confused the authors themselves and led to controversy.

8.5 Gregory on impossibility proofs and the new analysis

We saw above that eighteenth-century mathematicians like Montucla and Condorcet stressed the necessity of analytic methods for the proof of the impossibility of the duplication of a cube and the trisection of an angle. The

[3] That 2a is false was proved by Liouville when he showed that the trigonometric functions are transcendental (see Chapter 14). That 2b is false follows from Lindemann–Hermite's theorem according to which v and $\sin v$ cannot be algebraic simultaneously if $v \neq 0$. Thus, constructible cords have transcendental arcs or sectors. Statement 2c is correct even if we exclude the obvious counterexample $v = 0$. Indeed for each rational value of q the equation $\sin v = v + q$ has a solution v (a different solution for each value of q). For these countably many values of v, the arc (or the sector) is algebraically (even rationally) expressible in terms of the chord.

Scottish mathematician James Gregory (1638–75) was probably the first who made this connection between the new analysis and proofs of impossibility. In the preface to his *Vera circuli et hyperbolae quadraturae* (1667) he wrote:

> But the task of analysis as well as common algebra is not only to solve problems but also (if this turns out to be the case) to prove their impossibility.
>
> (Gregory 1667, 408)

Like Simplicius before him, Gregory pointed to the similarity of the problem of the quadrature of a circle and the incommensurability of the side and diagonal in a square, but in contrast to Simplicius he asserted the impossibility of the first problem:

> For it is the same thing (as I will demonstrate in this treatise) to exhibit the ratio of the circle to the square of its diameter analytically or in a form hitherto known, as to exhibit the ratio between the side of a square and its diagonal in a commensurable way.
>
> (Gregory 1667, 410)

As we shall see in Section 8.6, he later insisted that he had indeed proved the impossibility of the algebraic quadrature of a circle, but in the Introduction he admitted that he had not given a complete proof in geometric language.

In a very remarkable continuation of the above quote, Gregory stressed the desirability of developing a general theory of analytic quantities and their incommensurables. He found it surprising that no one had ever written such a work (except for Euclid's *Elements* Book X) since in this vast area one could prove not only when geometric problems could be solved geometrically or analytically or when they needed recourse to mechanical curves, but also why constructions using the mesolabium of Eratostenes (that constructs mean proportionals or nth roots) cannot always be replaced by constructions by ruler and compass. Moreover such a theory would show when equations can be reduced to pure equations (extraction of radicals). According to Gregory such a treatise would "not only be useful for speculative geometry but also very admirable" (Gregory 1667, 10–11).

This may be the first time in the history of mathematics where impossibility theorems were generally mentioned as an important field of inquiry that could be pursued by mathematical methods. It is particularly noteworthy that Gregory wanted to investigate when equations can be solved by radicals. Here he was much ahead of his contemporaries and immediate successors, for whom the problem was a constructive search for the solution of the equation. Only with Abel and Galois around 1830 do we find an attempt to answer Gregory's question. In a more general way, Gregory's call for a theory of incommensurable quantities may point in the direction of the later theory of transcendental numbers and functions.

8.6 Gregory's argument for the impossibility of the algebraic indefinite circle quadrature

Gregory's proof of the impossibility of the definite circle quadrature mentioned in the introduction to his *Vera circuli et hyperbolae quadraturae* was only a corollary to a proof of the impossibility of the indefinite algebraic quadrature of a circle.[4] His line of attack was to examine an approximation method similar to that of Archimedes', and from its structure to conclude that no algebraic formula could express the area of a sector exactly in terms of its cord.

He explicitly formulated a concept of a quantity "composed" of other quantities through any number of imaginable operations. Gregory called such a composed quantity analytic if it is composed of the other quantities using only the five analytic operations: addition, subtraction, multiplication, division, and extraction of roots. Heinrich (1901) stated that Gregory's concept of a quantity composed of other quantities is equivalent to Euler's later function concept. However, I think there is a difference between the two concepts, a difference that will be important in the following: Where Euler's function of a variable quantity is an analytical expression composed of this variable and numbers or constant quantities, there are no such numbers or constant quantities in Gregory's formulation. That means that in Gregory's analytic "composits" rational numbers may enter (for example, 2/3 can be written $(x + x)/(x + x + x)$, where x is one of the quantities of which the composit is composed) as well as explicit algebraic numbers. But transcendental numbers cannot enter. That means that where πx is an algebraic (even rational or integer) function of x for Euler, it is not an analytic "composit" in Gregory's sense. Moreover, Gregory's analytic "composits" are explicit expressions, whereas Euler's analytic functions can be given implicitly through an equation. In the following I shall write $f(x, y)$ when Gregory would write "a quantity composed analytically of x and y."

[4] Gregory's argument is analyzed in Dehn and Hellinger (1939), Scriba (1983), and Crippa (2019). See also Gregory (1939).

As Archimedes had done in his *Measurement of the Circle*, Gregory approximated the circle by inscribed and circumscribed polygons and investigated what happens when the number of sides of these approximating polygons is successively doubled. But in contrast to Archimedes he considered the areas rather than the circumferences of the figures, as well as considering an arbitrary sector of the circle rather than the whole circle.

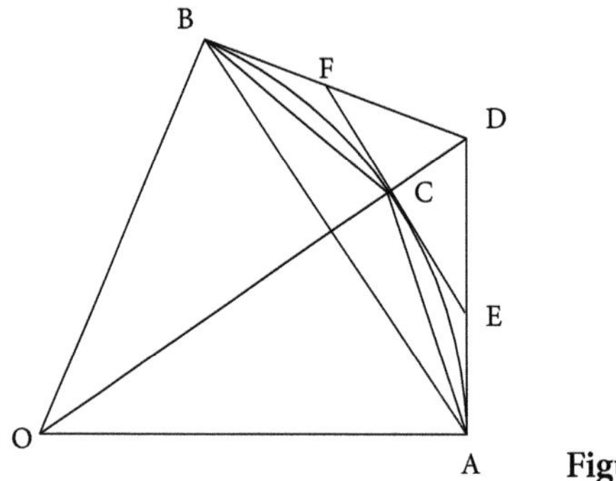

Figure 8.4 Gregory's procedure

Let OAB be a sector of a circle (Figure 8.4). The first inscribed polygon I_1 is the triangle OAB. The second inscribed polygon I_2 is the quadrilateral $OACB$ where C is the midpoint of the arc AB. The third inscribed polygon I_3 is similarly obtained by bisecting each of the arcs AC and BC, etc. The first circumscribed polygon C_1 is the quadrilateral $OADB$ where AD and BD are tangents to the circle, and the second C_2 is the pentagon $OAEFB$ where EF is the tangent of the circle through C, etc. In propositions I–VI Gregory proved that

$$I_{n+1} = \sqrt{C_n I_n}, \quad C_{n+1} = \frac{2C_n I_{n+1}}{C_n + I_{n+1}} = \frac{2C_n I_n}{I_n + \sqrt{C_n I_n}}. \tag{8.7}$$

This double series of inscribed and circumscribed polygons form what Gregory called a convergent series (the origin of this term). Since the area S of the sector lies between I_n and C_n for all n Gregory concluded that the limit ("termination" in his formulation) of the series must be equal to S. He considered S as a composite of I_1 and C_1, and thought of the operation that determined S from I_1 and C_1 as a sixth type of operation in addition to the five analytic or algebraic operations. He wanted to prove that this operation was not analytic.

In order to achieve this goal, Gregory pointed out that if one can find an analytic (algebraic) composite f of two variables such that

$$f(I_n, C_n) = f(I_{n+1}, C_{n+1}) \tag{8.8}$$

and if K denotes the constant value of $f(I_n, C_n)$ and we let n tend to infinity, we get

$$f(S, S) = K = f(I_1, C_1). \tag{8.9}$$

This is an algebraic equation from which S can be determined as a composite[5] of I_1 and C_1:

$$S = F(I_1, C_1). \tag{8.10}$$

Gregory gave an example of a different converging double sequence where this method could be used to find the limit, but then went on to argue that in the case of the double sequence of inscribed and circumscribed polygons, a composite f satisfying (8.8) does not exist.

In order to carry through this argument he introduced two new parameters u, v such that

$$I_1 = u^2(u+v) \quad \textit{and} \quad C_1 = v^2(u+v). \tag{8.11}$$

It is not hard to see that conversely u and v can be expressed as algebraic composites of I_1 and C_1. Thus, the area of the sector S is an algebraic composite of I_1 and C_1 if and only if it is an algebraic composite of u and v. From the recursion relations (8.7) Gregory concluded that

$$I_2 = uv(u+v) \quad \textit{and} \quad C_2 = 2uv^2. \tag{8.12}$$

Inserting the expressions (8.11) and (8.12) into (8.8) Gregory concluded that the composite f must satisfy the identity

$$f(u^2(u+v), v^2(u+v)) = f(uv(u+v), 2uv^2), \tag{8.13}$$

He claimed that such an algebraic composite f could not exist[6] and gave two arguments for it:

[5] Here Gregory seems to imply that S can be found as an explicit algebraic "composite," which does not agree with his statement that general algebraic equations may not be solvable by radicals.

[6] In Huygens' words: "il n'y a aucune quantité qui puisse estre composée analytiquement & de mesme manière, des termes $a^3 + aab$, $abb + b^3$ & des termes $aab + bba$, $2bba$" (Huygens 1668a, 229). As we shall see in Section 8.7, Huygens did not believe Gregory's statement.

1. As a result of the u^3 term, the left-hand side, $f(u^2(u+v), v^2(u+v))$, will always contain higher powers of u than the right-hand side, $f(uv(u+v), 2uv^2)$, which only contains u to the second power.
2. $f(u^2(u+v), v^2(u+v))$ will contain more terms than $f(uv(u+v), 2uv^2)$.

Hence, Gregory concluded that one cannot express the area of a sector as an algebraic composite of its inscribed triangle and its circumscribed quadrangle.

It is easy to see that the inscribed triangle I_1 and the circumscribed quadrangle C_1 are algebraic functions of the radius r of the circle and the chord $c = AB$ of the sector and the converse is also true. The question of the indefinite algebraic circle quadrature as formulated in Section 8.4 is thus equivalent to the question whether the area of the sector is an algebraic composite of I_1 and C_1. Gregory therefore claimed to have proven the impossibility of the indefinite algebraic circle quadrature.

8.7 Huygens' and Wallis' critique of Gregory

Only one year after the publication of Gregory's book, the Dutch mathematician Christiaan Huygens (Figure 8.5) publicly challenged Gregory's impossibility argument in a review in the *Journal des Sçavans* (Huygens 1668a, 228–30). The review soon led to an extensive correspondence (Huygens 1895, 228–399; Huygens 1940, 259–315) between Huygens and his fellow members of the Royal Society. Huygens' first objection was that Gregory had not given sufficient demonstration of the impossibility of finding an algebraic function f satisfying (8.13). Huygens' counterexample was incorrect (see Heinrich 1901, 84) but Gregory's repeated attempts to repair his own original unconvincing arguments never came much closer to a proof.[7]

Second, Huygens pointed out that even if Gregory succeeded in proving that an algebraic composite f satisfying (8.13) does not exist this would only prove that the indefinite algebraic circle quadrature could not be obtained using the successive method suggested by Gregory. Huygens maintained that other methods might still lead to the goal. Indeed, Gregory's initial argument is vulnerable to this objection and also in this case he tried to repair the problem. His counterargument was directed to the president of the Royal Society, Henry Oldenburg, and published in its *Philosophical Transactions* of July 13, 1668 (Gregory 1668b, 240–3). In this first defense of his methods Gregory tried to argue that if the area can be expressed algebraically as a function F

[7] For a closer scrutiny of the mathematical arguments see Lützen (2014) and Crippa (2019).

Figure 8.5 Christiaan Huygens (1629–95)
Reproduced by Jim Høyer DMS

of the inscribed triangle and the circumscribed quadrangle as in (8.10) then this function must itself necessarily satisfy the identity

$$F(I_n, C_n) = F(I_{n+1}, C_{n+1}), \tag{8.14}$$

which is of the form (8.8). Thus, if the indefinite quadrature of a circle is possible, then Gregory's method would lead to it and Huygens' objection would be invalid.

Gregory's argument for this claim is strange and did not convince Huygens. Yet, the result is, in fact, correct[8] if it is assumed that the same composite F works for all sectors. Thus if Gregory intended to prove the impossibility of the indefinite circle quadrature in the sense of 2a from Section 8.4, this step in the argument is indeed correct, but his argument for it is not.

I have already mentioned that Gregory in the introduction to his *Vera Circuli*... claimed to have demonstrated that the definite circle quadrature is impossible algebraically. However, this theorem was nowhere formulated or proved explicitly in Gregory's book, and as we have also seen, he admitted that he had not provided a proof in geometric language. Thus it is not surprising that an attentive reader such as Wallis at first did not get the impression that Gregory claimed to have proven it (Wallis 1668, 283). Yet, Gregory reaffirmed in later communications that this was his intent. His argument seems to have been that since he had shown in general that any sector could not be squared

[8] For a proof quite different from Gregory's, see Lützen (2014, 229–30).

algebraically, then in particular the quarter circle (and thereby the full circle) cannot be squared algebraically. Huygens, in his critique of Gregory's book, pointed out that this conclusion is invalid, and in a report written in the fall of 1668 to Lord Brouncker (1620–84) concerning the discussion between Gregory and Huygens, Wallis agreed with Huygens (Wallis 1668). He pointed out that one can show that the trisection of an angle is impossible in general,[9] but that there are nevertheless special trisectable angles such as the right angle. Similarly, he argued, the indefinite circle quadrature can be impossible while some cases (as, for example, the entire circle) could be analytically squared.

The reason for the disagreement seems to be at least partially due to the ambiguities pointed out above. Indeed, Gregory's own formulation is inherently unclear:

> Prop. XI. Theorem: I say that the circular, elliptic or hyperbolic sector ABIP is not analytically composed from the triangle ABP and the trapezoid ABFP.[10]

This can be interpreted as any of the alternatives $\neg 2a$, $\neg 2b$, or $\neg 2c$. If Gregory thought that he had proved $\neg 2c$ he would, in fact, be able to conclude that the definite circle quadrature was analytically impossible. However, as pointed out in Section 8.4, $\neg 2c$ is blatantly false so it is understandable that Huygens and Wallis read Gregory as though he claimed $\neg 2a$ or $\neg 2b$. In both cases, they were right in pointing out that the impossibility of the definite circle quadrature does not follow as a consequence.

Huygens' reading of Gregory seems to be close to impossibility statement $2a$ in Section 8.4:

> Thus in order to conclude that the ratio of the circle to the square of its diameter is not analytic, one needs to demonstrate not only that the sector of the circle is not indefinitely analytical ... but that it is also true in all definite cases.[11]

[9] As we saw in Section 7.3, Descartes claimed that he had proven this. Wallis' formulation of the impossibility of the indefinite angle trisection is strange. Having written the cubic equation determining the chord of the third of an angle in terms of the chord of the angle itself, he wrote that this solution "cannot be universally designed by those he (Gregory) calls Analyticall operations" (Wallis 1668, 285), and he referred to Charles (perhaps Descartes?) and Schoten. In view of Cardano's formulas this may sound surprising, but perhaps he was referring to the unavoidability of complex numbers in this irreducible case. He nowhere mentioned ruler and compass.

[10] "Prop. XI. Theorema: Dico sectorum circuli ellipseos vel hyperbolae ABIP non esse compositum analyticè à triangulo ABP & trapezio ABFP" (Gregory 1667, 419). Gregory dealt with all the three conic sections at once.

[11] "Pour conclure donc que la raison du Cercle au Quarré de son diametre n'est pas analytique, il falloit demontrer non seulement que le Secteur de cercle n'est pas analytique indefinitè ...; mais que cela est vray aussi *in omni casu definito*" (Huygens 1668b, 273).

Wallis on the other hand seems to have believed that formulation 2*a* and 2*b* were equivalent:

> That his 11th proposition,[12] though ever so well demonstrated, shews onely yt ye Sector indefinitely considered can not be so compounded as is there sayd: Or, (which is equivalent) not every Sector. Notwithstanding which it might well inough be possible, that some Sector (if not all) might be analyticall to its Triangle or Trapezium: (And I think he doth allow it so to bee, or even commensurable ...).
>
> (Wallis 1668, 284)

The quote shows Wallis' uncertainty about the exact meaning of Gregory's statements, and the entire correspondence shows that it was difficult for the correspondents to formulate the subtle differences between the different impossibility statements. In particular, this seems to explain the differences in opinion about the relation between the impossibility of the definite circle quadrature and the indefinite one.

Overall, Wallis agreed with Huygens as far as the indefinite circle quadrature is concerned. On the definite quadrature they disagreed. When Wallis discovered that Gregory claimed to have proved its impossibility, Wallis not only rejected Gregory's proof, but suddenly claimed priority for the result, interpreting his reflections in the *Arithmetica Infinitorum* as a proof of impossibility. Huygens, on the other hand, bought neither "proof" and maintained that it was still an open question whether π was algebraic or even rational.

8.8 Leibniz on the impossibility of the indefinite circle quadrature

The two "inventors" of differential calculus, Newton and Leibniz, also came up with arguments for the impossibility of the algebraic indefinite circle quadrature. In contrast to Gregory's argument their arguments can be made rigorous from a modern point of view.

Crippa (2019, 97, 138) has shown that Leibniz (Figure 8.6) heard of Gregory's dispute with Huygens when he lived in Paris and was mentored by the

[12] Wallis accepted Gregory's argument that no function satisfying (8.13) exists, but like Huygens he argued that that only means "that ye converging series cannot his way be determined, not that it can no way be determined analytically" (Wallis 1668, 284). So, in fact, he did not accept the conclusion of Proposition XI.

latter in modern mathematics and science. It inspired him to a deep study of the quadrature of a circle that resulted in his discovery of his famous series (8.2) for π. This research was intimately connected with his invention of calculus (see, e.g., Crippa 2019, 93–146). In 1675–6 Leibniz composed a manuscript on the matter entitled *Preface to a small work on the arithmetic quadrature of the circle* but it was not published until 1993 by Eberhard Knobloch (Leibniz 1675/6). Leibniz shared Huygens' critical attitude toward Gregory's impossibility proof but came up with a proof of his own.

Figure 8.6 Gottfried Wilhelm Leibniz (1646–1716)
Reproduced by Jim Høyer DMS

Leibniz's argument is brief, clear and simple (Leibniz 1675/6, 7). It is a proof by contradiction: Assume that the indefinite quadrature (or equivalently the indefinite rectification) of the circle was possible algebraically, i.e., that there were an algebraic equation

$$P(\sin(v), v) = 0 \tag{8.15}$$

relating the arc v and its sine where P is a polynomial of degree m.

> Under this assumption one can draw a curve of the same degree so that when the abscissa expresses the sine, the ordinate will express the arc and conversely. Thus using this curve one can divide a given angle or arc in a given ratio, or find the sine of an arc that has a given ratio to a given arc. Thus the problem of the universal division of the angle will be of a definite degree.
>
> (Leibniz 1675/6, 7)

Here Leibniz referred to the problem of dividing an angle in a given ratio. This generalization of the classical problem of trisection of an angle asks for the construction of an angle which is $^m/_n$ of a given angle. Since the addition of angles is a simple matter, the problem boils down to dividing an angle into n parts.

However, according to Leibniz it was known that the division of an angle into n parts depends on an equation of degree n (at least when n is odd). But according to Leibniz this contradicts that the problem can be solved by an equation of a definite degree independent of n.

Leibniz was much clearer than Gregory about what he believed he had shown:

> But to present the relation between the arc to [its] sine in general by an equation of a certain degree is impossible.
>
> (Leibniz 1675/6, 6)

Even clearer is the quote

> Except that through this rule [Leibniz's series for Arctan] not only the whole circle but also an arbitrary part of it, and not only the whole circumference but also any arc of it can be found, which is impossible by a fixed analytic expression.
>
> (Leibniz, 1675/6, 5)

As formulated here the impossibility statement clearly corresponds to $\neg 2a$ from Section 8.4 except Leibniz took the radius of the circle to be equal to 1 and formulated his impossibility primarily in terms of the rectification rather than the area. Moreover, as opposed to Gregory he did not insist on an explicit algebraic expression in terms of radicals, but simply asked whether there could be an algebraic equation relating the arc and its sine. Thus, as formulated by Leibniz, the impossibility statement undoubtedly says that sin (or equivalently arcsin) is not an algebraic function.

Leibniz also quite clearly stressed that this impossibility of what he called the full quadrature (Quadratura plena) (what we have called the indefinite quadrature) did not imply the impossibility of the definite algebraic circle quadrature:

> That this is impossible (the algebraic definite circle quadrature) was asserted by the very ingenious Scotsman Gregory in his book De Vera Circuli

> Quadratura, but he has not provided a proof there, if I am not mistaken. I still do not see what prevents the circumference itself or a determined part of it to be measured, and that the ratio of a certain arc to its sine can be expressed by an equation of a certain degree.
>
> (Leibniz 1675/6, 6)

In other words Leibniz agreed with Huygens that it had not been proved that π was transcendental.

From a modern point of view, the only problematic point in Leibniz's argument is his claim that the division of the angle into n parts requires an equation of degree n. To be sure, Viète had derived such an equation when n is odd, but in order to get to a contradiction Leibniz would need to show that no equation of lower degree will do the job, or at least that there is no upper bound to the degree of these equations. That would require an irreducibility argument that Leibniz did not give.

Even though Leibniz's impossibility argument was not published until recently, the impossibility result was mentioned in many of Leibniz's publications (see, e.g., Leibniz 1858, 92, 120, 124). In the paper "De dimensionibus figuram inveniendis" that was published in *Acta Eruditorum* in 1684 he mentioned that the impossibility of the indefinite algebraic circle quadrature "can be proved easily in many ways" (Leibniz 1684, 124), but he did not reveal any of the ways. He emphasized that this impossibility did not imply that the definite circle quadrature was algebraically impossible as well. In order to bring out this non-sequitur he gave an example of an algebraic curve whose quadratrix (area curve) was non-algebraic but whose total area was algebraic. However, he only sketched his arguments for these claims. He claimed that if there were an algebraic quadratrix he could prove that it must be of a particular form, and he further argued that no curve of this form would do as a quadratrix (Leibniz 1684, 125–6).

8.9 Newton's argument for the impossibility of the algebraic indefinite oval quadrature

In Lemma 28 of Book 1 of the *Principia* (1687) Isaac Newton (Figure 8.7) proved that an arbitrary oval cannot be squared indefinitely analytically. This theorem has the impossibility of the indefinite algebraic quadrature of a circle as a simple corollary. Newton formulated his theorem as follows:

No oval figure exists whose area, cut off by straight lines at will, can in general be found by means of equations finite in the number of their terms and dimensions. (Newton 1687, 511)

Figure 8.7 Isaac Newton (1643–1727)
Reproduced by Jim Høyer

Newton's theorem and its proof were discussed at great length by his contemporaries as well as by modern mathematicians and historians of mathematics. Arnol'd has given a mathematically exhaustive discussion of Newton's theorem in the book *Huygens and Barrow, Newton and Hooke* (Arnol'd 1990) and in the paper "Newton's Principia Read 300 Years Later" (Arnol'd and Vasil'ev 1989); Pourciau (2001) and Pesic (2001) have dealt with the matter in a more historical fashion. The following discussion is based on their work.

Newton did not specify explicitly what he meant by an oval, but it is rather clear that it must be a closed convex algebraic curve.

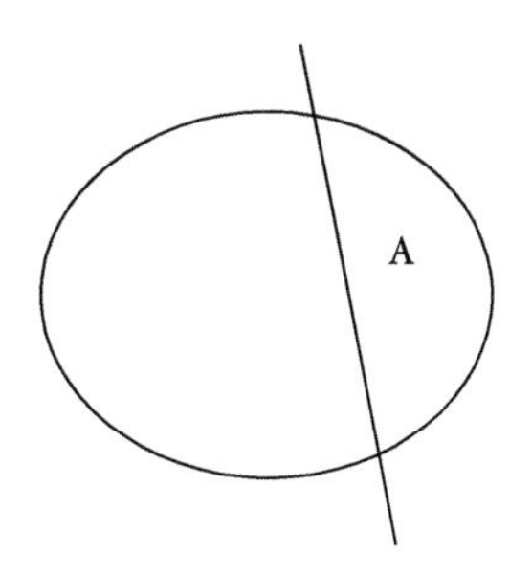

Figure 8.8 The area *A* cut off by a line

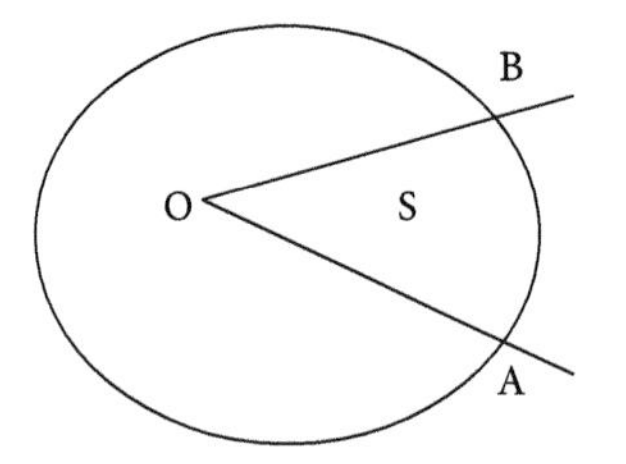

Figure 8.9 A sector S cut off by a fixed line OA and a line OB revolving around O

From the formulation of Newton's theorem and his subsequent proof we can infer that the theorem was intended to say that there is no polynomial equation $P(A, a, b, c)$ that determines the area A in a given oval cut off by an arbitrary line, where a, b, c are the coefficients of the equation of the line (Figure 8.8).[13] Newton selected an arbitrary point O inside the oval and a fixed half-line OA that intersects the oval at A (Figure 8.9). We can reformulate the claim of the theorem into the following equivalent claim: If S denotes the sector AOB cut out of the oval between OA and another (variable) line OB (B on the oval), then there is no algebraic relation between S and B, or more precisely, there exist no polynomial equation

$$P(S, x) = 0 \tag{8.16}$$

relating S and the x-coordinate of B.[14] In a slightly modernized way we can formulate the idea of Newton's argument as follows: If $\bar{A}$ designates the area of the entire oval and if S is the sector corresponding to a particular point $B = (x, y)$, then $P(S + n\bar{A}, x) = 0$ for all natural numbers n. Indeed, if we let OB rotate one full rotation, B and thus x will return to its former value but the area S will have increased by $\bar{A}$. After n rotations x will return to the same value and the area will have increased by $n\bar{A}$. But that means that Eq. (8.16), considered as an equation in S for fixed x, will have infinitely many solutions namely $S + n\bar{A}$, for any natural n. However, this is not possible if P is a polynomial of finite degree.

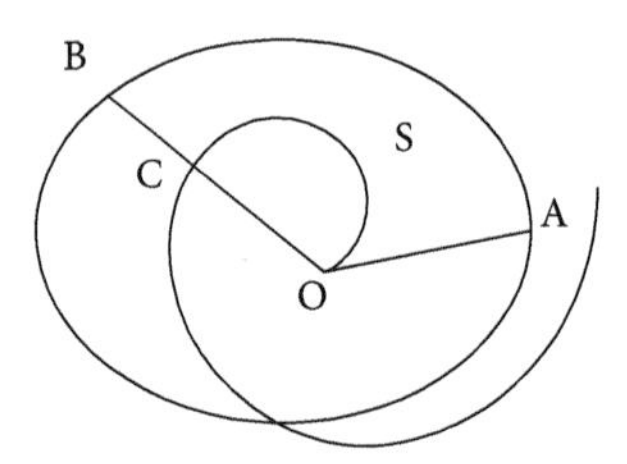

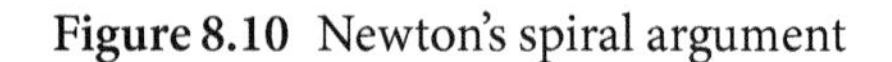

Figure 8.10 Newton's spiral argument

[13] In his formulation Newton also seems to include the coefficients of the equation of the oval, but since they are constants we can leave them out.

[14] Or equivalently the cord AB.

Newton formulated the proof somewhat differently, probably in order to bring out the multivaluedness more clearly, and in order to formulate the contradiction as a geometric contradiction concerning intersections of curves. He generated a new spiral-shaped curve (Figure 8.10) as the trajectory of a point C moving on the half-line OB while this half-line rotates around O in such a way that the distance OC is everywhere proportional to (say equal to) the area of the sector AOB swept out in the oval by the half-line OB. Newton correctly remarked that if the area of the sector can be found "by means of a finite equation," then "all the points of the spiral can be found by means of a finite equation"; i.e., it is an algebraic curve. And thus "the intersection of any straight line, given in position, with the spiral can also be found by means of a finite equation" (Newton 1687, 511). But it is clear that any straight line (in particularly a line through O) will intersect the spiral in infinitely many points. And that cannot be, because an algebraic equation can only have finitely many roots.

Newton's theorem and argument has been hotly debated both by his contemporaries and by modern historians and mathematicians. Some of Newton's contemporaries, such as Huygens and Leibniz, considered the theorem to be incorrect, and gave various counterexamples. The counterexamples were either non-algebraic curves (such as a triangle, mentioned by Huygens) or algebraic curves with self-intersections as the lemniscate (mentioned by Leibniz).

Still, of the proofs put forward by seventeenth-century mathematicians for the impossibility of the algebraic indefinite circle quadrature, Newton's stand out. As Leibniz's proof, it can be made correct, but in contrast to Leibniz's proof it was published at the time, even in a prominent place. Its striking generality and qualitative nature also set Newton's theorem and its proof apart from the other seventeenth-century arguments. And despite the criticisms put forward by some of Newton's contemporaries and immediate successors, Newton's argument was widely accepted by the mathematicians of the eighteenth century (see Chapter 9 and Whiteside's note 126, p. 306 of Newton 1974).

8.10 Why prove impossibility

We saw in Section 8.5 that Gregory emphasized the importance of proving impossibility results. Still, it is conspicuous that all the impossibility arguments we have dealt with in this chapter were formulated in order to highlight a positive result concerning the quadrature of a circle. This is most obvious in

Leibniz's case. Leibniz's series (8.2) was a special case of what he called the full quadrature (and what we have called the indefinite quadrature)

$$\frac{b}{1} - \frac{b^3}{3} + \frac{b^5}{5} - \frac{b^7}{7} + \cdots, \tag{8.17}$$

expressing the arc in a unit circle (less than a quarter circle) with tangent equal to b. The impossibility proof of the indefinite circle quadrature served the purpose of proving the "corollary": "There cannot be found a more perfect full analytic quadrature whose dimensions of the terms are rational numbers, than the one we have given" (referring to the series (8.17); Leibniz 1675/6, 6).

In other words, the purpose of the impossibility proof was to show that Leibniz's own solution to the problem of the quadrature of a circle was the simplest possible at least in the case of the indefinite (full) quadrature. In the case of the definite quadrature Leibniz was less assertive. As pointed out above, he admitted that it had not been proven that this problem could not be solved by a finite equation or even by rational numbers. However, he argued that even if it turned out to be possible to find such a solution, it was doubtful that it could be as beautiful as his series (8.2) that possesses a "wonderful simplicity" (Leibniz 1675/6, 5–7).

In a slightly less obvious way Gregory's and Newton's impossibility results exemplify the same thing. The "true circle quadrature" that the title of Gregory's book refers to is a very accurate approximation procedure that he had discovered and that he published in the book. In fact, in the correspondence between Gregory, Huygens, and their colleagues this constructive procedure was discussed in a more aggressive tone than the impossibility result. Huygens directly accused Gregory of plagiarism, claiming that the Scottish geometer had copied a method that he had himself discovered previously and communicated to the Royal Society. So also in Gregory's book the impossibility result served the purpose of showing that one could not hope for a better solution of the indefinite circle quadrature than the approximate one found by Gregory himself.

Newton formulated his impossibility result in connection with the solution of the Kepler problem. The problem is to determine the position of a planet at a given time from Kepler's laws. The first law states that the planet rotates around the Sun in an ellipse with the Sun in one of the foci and the second law states that the area traversed by the radius vector from the Sun to the planet is proportional to time. Thus, when the time is given, the area of the appropriate sector of the ellipse is known and the problem is to determine the position of the planet from this area. Newton's impossibility result shows that the rectangular coordinates of the planet cannot be determined from the area (and thus from the time) by way of a polynomial equation or by the use of a geometric rational curve. "Therefore I cut off an area of an ellipse proportional to the time by a geometrically irrational curve as follows," Newton continued (Newton 1687, 513) and then went on to present his own solution of the Kepler problem. Again we see that the impossibility problem did not stand alone but was an integral part in an argument showing that a certain constructive solution of a problem was the simplest because a simpler one did not exist.

As pointed out by Guicciardini, Newton's impossibility result can be seen in a wider context as one among several arguments put forward by the mature Newton to show the inadequacy of Descartes' analytic methods (Guicciardini 2009, 305–8).

9
Circle Quadrature in the Eighteenth Century

During the eighteenth century the problem of the quadrature of the circle became widely known also among amateur mathematicians. Many tried to solve the problem, and in order to gain eternal fame or a supposed prize, many sent their solutions to the French Academy. In the end (1775), the academicians were fed up evaluating the works of circle squarers and decided to stop reviewing circle quadratures, cube doublings, angle trisections, and designs of perpetual motion machines.

Some amateurs and learned non-mathematicians such as Georges-Louis Leclerc, Comte de Buffon (1707–88), and Joseph Panckoucke (1736–98) tried to prove the impossibility with metaphysical arguments but their arguments were not accepted by the academy either.[1] At least two authors rediscovered Leibniz's argument for the impossibility of the indefinite quadrature of the circle (apparently independently of Leibniz). Let us have a look at those relatively sound arguments.

9.1 Joseph Saurin (1659–1737)

The two rediscoveries of Leibniz's argument were published by the French Academy of Sciences in their *Histoire de l'Académie royale des Sciences* for the years 1720 and 1727 (published 1722 and 1729, respectively). The first one was a very clear presentation by Saurin entitled *Démonstration de l'impossibilité de la quadrature indefinie du Cercle, Avec une manière simple de trouver une suite de droites qui approchent de plus en plus d'un Arc de Cercle proposé, tant en dessus qu'en dessous* (Saurin 1722). Like the earlier works we have dealt with, the paper combined an impossibility proof with an approximate constructive method for solving the problem, but in contrast to

[1] Jeff Loveland (2004) has written a highly commendable paper on the interaction between amateur circle squarers and professional mathematicians in 18th-century France and Panckoucke in particular.

A History of Mathematical Impossibility. Jesper Lützen, Oxford University Press.
 DOI: 10.1093/oso/9780192867391.003.0009

the earlier works, the impossibility proof was mentioned first in the title. This may indicate that this part of the question had gained importance.

Saurin first focused on the problem of dividing a circular arc in a given ratio. He proved that the various curves that had been introduced in order to solve this problem such as Archimedes' spiral, the quadratrix, and the cycloid, and a new curve he introduced in the paper must necessarily be mechanical (transcendental). Since he phrased his argument in terms of the new curve, I shall introduce it here. This curve (see Figure 9.1) is generated by the intersection point *M* of two straight lines, a horizontal line *NM* which moves parallel to itself while the point *N* moves uniformly along the quarter circle *AB* from *A* to *B* and a vertical line *QM* which moves parallel to itself while the point *Q* moves uniformly along the line *CD* from *C* to a given point *D*.

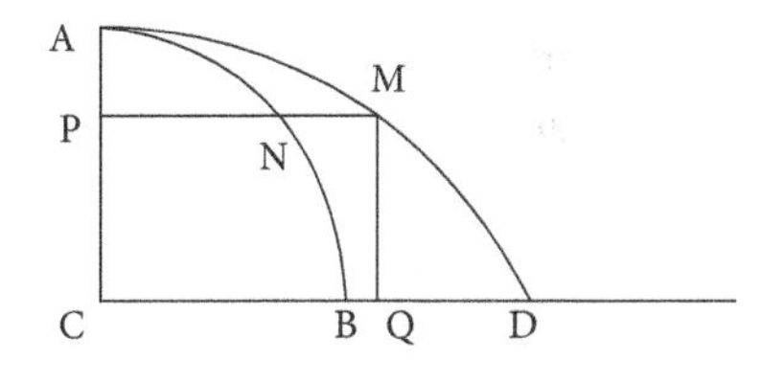

Figure 9.1 Saurin's argument

This curve can be used to divide a given arc in a given ratio; in particular, one can divide an arc into *n* parts. Indeed one only needs to find the line segment *QQ'* corresponding to the given arc *NN'* and divide it into *n* equal parts and then go back up to the circular arc. Saurin then referred to book 10 of l'Hospital's *Traité analytique des sections coniques* (1720) article 443 where l'Hospital had proved that the degree of the equation that one must solve in order to divide an arc into *n* parts must increase to infinity as the number of parts increase. Saurin also referred to Chapter 9 of the same book where it was proved that if an equation is solved by the intersection of two algebraic curves and if one of these curves remain the same, then the degree of the second must increase toward infinity if the degree of the equation increases to infinity. But since the curve *AMD* can be used to divide any arc into any number of equal parts by one and the same method (using only straight lines and circles), this curve cannot have a fixed degree, and must thus be mechanical or transcendental.

Now Saurin turned to the indefinite circle quadrature, or rather the equivalent question of the indefinite circle rectification. If this problem could be solved algebraically, there would be an algebraic relation between the ordinate *AP* and the corresponding arc *AN*. Thus, if we describe a curve having *AP* as its ordinate and its abscissa *PM* equal to the corresponding arc *AN*, this curve would be algebraic. Saurin also remarked that this curve is one of the curves mentioned above, namely when the given point *D* is such that *CD* is equal to the quarter arc *AB*. Saurin calls this curve the "ligne des arcs." But by the angle division argument above, this curve cannot be a geometric (i.e., algebraic) curve, so Saurin concluded that the algebraic circle rectification is impossible.

9.2 Anonymous

As indicated by its title the *Histoire de l'Académie Royale des Sciences Avec les Mémoires de Mathématique & de Phisique, pour la même Année* was divided into two parts: The *Histoire* containing anonymous official papers or statements from the academy on the state of affairs in particular areas of science and the *Mémoires*. Saurin's paper was contained among the *Mémoires*. In the same volume of the journal one can find in the *Histoire* section a paper entitled *Sur la Rectification indefinie des Arc de Circle* (Anonymous 1722). In contrast to Saurin's paper the anonymous paper dealt mostly with the definite quadrature or rather rectification of the circle and in contrast to Saurin's clear exposition, the anonymous paper was somewhat rambling and imprecise. The treatment of the definite rectification concluded that it is possible that the rectification of the circle is "absolutely impossible," but "it is impossible, to demonstrate that one is in this case even when one is in it" (Anonymous 1722, 57). The anonymous author dealt with the indefinite circle quadrature in a way similar to Saurin and acknowledged the similarity (Anonymous 1722, 63).

9.3 Thomas Fantet De Lagny (1660–1734)

Seven years later a similar impossibility argument was apparently independently put forward by De Lagny. According to the biography by Pierre Costabel (2008) Lagny was a skilled mathematician who specialized in computational mathematics. For example, he calculated the value of π to 120 decimals. His impossibility argument was included as a comment to some of his approximations to trigonometric functions in the *Troisième Mémoire sur la Goniometrie purement analytique* published in 1729 in the *Histoire de l'Académie Royale des Sciences* (for the year 1727) (Lagny 1729). The argument was in essence the same as the one put forward by Leibniz and Saurin, but in contrast to those two arguments, it did not involve geometric curves. This feature, which gives the argument a more modern appearance, must have been natural to a calculator like Lagny and it is in agreement with the "purely analytic" nature of the paper highlighted in its title.

If t denotes the tangent of an arc x in a circle with radius r, then Lagny claimed that x/t and x/r cannot be determined algebraically from t/r or vice versa. In his demonstration, he first pointed out that the determination of the tangent of the nth part of an arc can be determined from the tangent of the arc itself by an equation of degree n.

> But if one could in general integrate the series represented by the exemplary formula above [giving the arc as a function of the tangent and the radius] by an equation of finite degree one would by one and the same equation or exact formula find the general ratio of the radius and the tangent and consequently the ratio composed of these two straight lines [the tangents corresponding to an arc and its nth part] to the arcs corresponding to their tangents and conversely, which was shown to be impossible. Thus neither the series above nor the exemplary formula that represents it can be integrated. Which was to be proved.
>
> (Lagny 1729, 125)

In a corollary Lagny further concluded that the determination of an arc in terms of its tangent and radius is impossible geometrically, because problems that can be solved geometrically can also be solved algebraically.

Lagny's formulation of the impossibility theorem clearly suggests that no arc can be expressed algebraically in terms of its tangent or conversely.[2] However, his proof clearly only proves that one cannot find one algebraic equation that determines an arbitrary arc in terms of its tangent, or equivalently arctan is a transcendental function. The concluding QED suggests that Lagny confused the two statements.

In the subsequent *Remarque* he gave a separate argument to show that the general impossibility implies the impossibility of the rectification of any specific arc. He based the argument on the "perfect uniformity of the curvature of the circle" from which he concluded the following:

> that the exact and geometric rectification of any particular arc of a circle is no more possible than the rectification of all other arcs in general. However, we have demonstrated the impossibility of this rectification in general. Thus, it is equally impossible for all particular arcs. For it is impossible to conceive

[2] The argument in Section 8.4, footnote 3 shows that this is incorrect in general, but correct for constructible arcs. However, Lagny's argument does not prove this.

> any reason for a difference in this respect between two arcs of the same circle of which one is supposed rectifiable and the other is not, not even any other arc which is not to this first arc as a number to a number.
>
> (Lagny 1729, 126)

As we saw earlier, such a symmetry argument had been rejected by Leibniz, and Lagny knew well that it was of a different nature than the previous and subsequent arguments in his paper. Still he tried to convince his readers that it was conclusive:

> This type of metaphysical and transcendental demonstration is in itself no less certain, no less convenient than those of the three following propositions which are nevertheless generally received and approved by all geometers as being well demonstrated.
>
> (Lagny 1729, 126)

The propositions he referred to are the following:

1. If one determines the value of the logarithm of a number,[3] for example, 10, in some arbitrary way it is impossible to determine exactly (in neither rational nor irrational numbers) the value of the logarithm of another number which is mutually prime with the first one.
2. The trisection of an angle cannot be solved by ruler and compass.
3. Two mean proportionals cannot be constructed by ruler and compass.

Lagny ended this section of his paper by concluding that the rectification of arcs of circles cannot be determined more simply, more promptly, or more exactly than through an infinite series that allows one to determine precise upper and lower bounds for its value. Thus, as most of his predecessors, he considered the positive solution as the main result, and the impossibility result merely served the purpose of showing that his own method of solution was the best possible.

9.4 The enlightened opinion

What was the general mid- to late-eighteenth-century opinion concerning the status of the impossibility of the quadrature of a circle? Given the varying

[3] "Nombre premier" seems to mean a first number rather than a prime.

opinions expressed by major mathematicians one may wonder whether it is meaningful to try to formulate a general opinion. However, the third quarter of the eighteenth century saw the appearance of a number of publications that can be considered as expressing an official view on the matter. These publications are:

1. Montucla's *Histoire des recherches sur la quadrature du cercle* from 1754;
2. D'Alembert's entry on the quadrature of a circle in the *Encyclopédie* (1765); and
3. The pronouncement written by Condorcet motivating why the Academy of Sciences would no longer investigate purported circle quadratures (1775/8).

According to Montucla, "the geometers today generally agree that the indefinite quadrature of the circle is impossible. However, as far as the definite quadrature is concerned they suspend their judgment" (Montucla 1754, 95). The belief in the impossibility of the indefinite circle quadrature was according to Montucla based on several proofs. After an approving account of Gregory's proof he admitted that it was not universally accepted. He attributed this lack of convincing power to the general nature of impossibility proofs:

> The negative demonstrations seem to have the shortcoming that they do not shed the same light as the positive ones and it is perhaps for this reason that those who have had as their goal to prove the impossibility of the quadrature of the circle, have never had a great success.
>
> (Montucla 1754, 93)

He raised no such doubts about Newton's proof. Finally, he referred to Saurin's proof and gave a summary of it.

Concerning the impossibility of the definite circle quadrature Montucla was less clear. As quoted above, he admitted that there was no general agreement about its possibility or impossibility, and he mentioned that there are curves that cannot be squared indefinitely but nevertheless have parts that can be squared. Here he referred not to Leibniz but to Bernoulli.

Yet Montucla claimed that the vain efforts of many mathematicians "seem to provide a proof of this impossibility which comes very close to a demonstration" (Montucla 1754, 101). In an "addition to Chapter III" at the end of the book, Montucla finally stated that upon reflection, he considered Gregory's argument to be a proof of the impossibility of the definite circle quadrature. His argument was based on the usual confusion about the meaning of the indefinite circle quadrature:

> for if it is true, as it seems one cannot disagree with him [Gregory], that in general the ratio of a segment [of an arc] or a sector to the inscribed or circumscribed polygon cannot be expressed by a finite function, then it is evident that this holds equally true of the entire circle and of any arbitrary particular sector.
>
> (Montucla 1754, 193)

This final afterthought was presented as Montucla's own idea, not as the generally accepted view. Thus, Montucla's authoritative book clearly stated the accepted view as follows: the impossibility of the indefinite algebraic circle quadrature had been demonstrated by valid proofs but there was no generally accepted valid proof of the impossibility of the definite algebraic circle quadrature.

9.5 D'Alembert

A similar view can be found in the article on the quadrature of a circle in the equally authoritative manifesto of the Enlightenment, the *Encyclopédie*. The author of the paper pointed out that the problem properly speaking consists in the alternative "to find this quadrature or prove it impossible." He continued proclaiming that

> M. Newton has already proved in the first book of his mathematical principles [*Principia*] ... that the indefinite quadrature of the circle and in general of all oval curves is impossible, that is that one cannot find a method for squaring at pleasure an arbitrary part of the area of the circle; but it is not yet proved that one cannot have the absolute quadrature of the entire circle.
>
> (D'Alembert 1765, 639)

This article is signed by (O) which reveals it as the work of d'Alembert (Figure 9.2). As in the case of Montucla, d'Alembert's statement of the official opinion was soon followed by a somewhat contrasting more personal opinion. Where Montucla had come to a varying opinion concerning the definite circle quadrature, d'Alembert cast doubt on the official opinion concerning the indefinite circle quadrature. This happened in a short note "Sur un autre paradoxe" in the fourth volume of his *Opuscules* (1768). Here he cast doubt on Newton's impossibility proof. Indeed, he pointed out that the arc length of a

Figure 9.2 Jean-Baptiste le Rond d'Alembert (1717–83)
Reproduced by Jim Høyer DMS

cycloid is an algebraic function of the distance to the baseline. More precisely if the x-axis is placed as in Figure 9.3, the arc from the origin to a point with coordinate x has length $2\sqrt{2rx}$ where r is the radius of the generating circle.

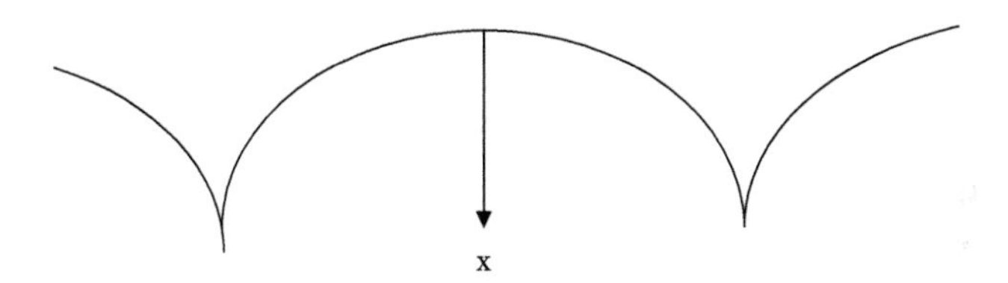

Figure 9.3 D'Alembert's objection to Newton's argument

However, d'Alembert pointed out that just as in Newton's argument for the impossibility of an algebraic oval quadrature, there are, in fact, infinitely many arcs corresponding to the same value of x, namely two values between each of the cusps. But according to Newton's argument, such an infinite valuedness should exclude an algebraic expression of the arc length. Of course, d'Alembert knew well why the argument fails: The expression of the arc length only holds until the point has reached a cusp. After a cusp, the arc length is expressed by a new algebraic function. But, d'Alembert asked, how do we know that a similar phenomenon does not hold in the case of an oval?

> It seems to me that these reflections could merit the attention of the geometers and engage them in searching for a more rigorous demonstration of the impossibility of the indefinite quadrature and rectification of oval curves.
>
> (d'Alembert 1768, 68)

With this interesting emphasis on the difference between the local and global algebraic rectification, d'Alembert called into question the rigor of Newton's argument, but he does not seem to have deprived it of all its demonstrative power. In particular he only seems to have considered the cycloid to be what Lakatos (1976) calls a local counterexample, i.e., a counterexample that problematizes a step in the proof without questioning the validity of the theorem itself. D'Alembert emphasized that the cycloid differs from an oval in that it is not a "courbe rentrante" (a closed curve) but he did not see that Newton's argument depended on that assumption. On the other hand, d'Alembert did not point out that the cycloid itself is not an algebraic curve as Newton's ovals are implicitly assumed to be.[4]

9.6 The French Academy of Sciences. Condorcet

Despite his doubts about the rigor of Newton's impossibility proof d'Alembert very strongly argued that the Academy of Sciences should refuse to evaluate circle quadratures, cube doublings, angle trisections, and perpetual motion machines submitted to the learned society (Jacob 2005). And when the decision was announced in 1778 the justification showed no doubt as to the impossibility of the indefinite circle quadrature:

> One can consider this solution [the rigorous circle quadrature] from two points of view: In fact one can search for the quadrature of the entire circle or the quadrature of an arbitrary sector whose chord is supposed to be known. The second of these problems is regarded as absolutely unsolvable. Grégori [*sic*], Newton, whose authority is so great even in a science where authority has so little power, have given different demonstrations of this indefinite quadrature. Jean Bernoulli has proved that the sector in question can be expressed by a real logarithm which however in its form contains imaginaries. It follows from this, that no real function be it algebraic or logarithmic and in a real form, can represent the value of an indefinite sector of a circle; that the equation between the sector and the cord cannot be constructed by the intersection of branches of curved surfaces be they real or rendered in a real form, and one can conclude from this reflection the absolute impossibility of the indefinite quadrature.[5]

[4] Arnold's completion of Newton's argument crucially depends on this assumption.
[5] Condorcet (1775/8, 63–4). In Montucla (1802, 641) the announcement is attributed to Condorcet.

On top of endorsing Newton's and Gregory's proofs the academy here introduced a new argument: Since the area (arcsine) can be expressed in terms of logarithms of an imaginary number,[6] it cannot be expressed as a real algebraic function. The somewhat surprising argument was not elaborated in the motivations of the academy.

9.7 Enlightening the amateurs

The authoritative writings by Montucla, d'Alembert, and the Paris Academy of Sciences give a clear picture of the general mid-eighteenth century opinion concerning the quadrature of a circle. The indefinite algebraic circle quadrature was believed to have been proved to be impossible by Newton and perhaps Gregory and others. The impossibility of the definite circle quadrature, on the other hand, was not believed to have found support in a rigorous proof. Yet, the belief in the impossibility of the latter problem was so strong that all the three authors strongly warned amateur mathematicians from wasting their time trying to square the circle. In fact, enlightening the general public and dissuading them from engaging in vain circle quadratures was a main goal of the three works studied in this section. Montucla displayed this purpose on the front page of his book where it was advertised as a "Work suitable for teaching the real discoveries made concerning this problem and for serving as a warning against new efforts to solve it" (Montucla 1754, front page).

In the *Encyclopédie,* d'Alembert referred the readers to Montucla's book where they could "be guarded against the promises, the boastings and the nonsense of circle squarers" (d'Alembert 1765, 639).

On behalf of the Academy of Sciences, Condorcet was more direct when he explained the motives behind the decision taken by the academy. In an often-quoted passage, he described how a great number of amateurs had ruined not only their own lives but also the lives of their families trying to square the circle and concluded:

> Thus humanity required that the Academy, who was persuaded of the absolute inutility of the investigation that it could have made of the quadrature of the circle, tried through a declaration to ruin the popular opinions that have been fatal to several families.
>
> (Condorcet 1775/8, 65)

[6] This result had been found by Johann Bernoulli in his correspondence with Euler that eventually led to Euler's explanation of the complex logarithm and his discovery of Euler's formulas (Katz 2009, 595–6).

These remarks indicate that the purpose of showing the impossibility of the quadrature of a circle had changed. Until the beginning of the eighteenth century the purpose of such impossibility proofs had been to set a positive solution of the problem into relief by showing that a simpler type of solution was impossible. This kind of purpose was absent from the three mid-eighteenth century works we have studied in this section. In these works, the impossibility proofs were presented as having a value in themselves, and the aim of publishing the impossibility was to enlighten humanity in order to prevent amateurs from wasting their own and their family's fortune trying to do the impossible. In this way, the impossibility results served a purpose central to the Enlightenment philosophers. However, as pointed out by Marie Jacob (2005) it is ironical that despite being inspired by Enlightenment ideals, the decision taken by the academy was, in fact, an authoritarian act. Though Montucla in the quote above declared that in mathematics authority played little importance, the academy had to admit that there was no rigorous proof of the impossibility of the algebraic definite circle quadrature, and so its decision was, at least in this respect, authoritarian in nature.

9.8 Lambert and the irrationality of π

In the German-speaking part of Europe, a similar warning to amateur mathematicians were formulated by Johann Heinrich Lambert (1728–77). In 1766 he wrote a *Preliminary information for those who seek the quadrature and the rectification of the circle*. It was published in 1770 as Chapter V of a book on various aspects of pure and applied mathematics (Lambert 1770). Like his French colleagues, Lambert tried to dissuade amateur mathematicians from wasting their time trying to square the circle. First, he disappointed their hopes of economic profit: The rumors about British and Dutch prizes for a successful circle quadrature were false. Second, he pointed out that contrary to common belief this problem had nothing to do with the very important problem of determining the longitude at sea. Finally, he argued mathematically that if the quadrature of a circle were possible, its solution would be difficult to find. In particular, he sketched a proof showing that that π is irrational.

Lambert was convinced that most circle squarers had not even grasped the elementary parts of geometry and so he doubted that they would even read his paper. However, even if Lambert's paper had little impact on amateur mathematicians, his proof of the irrationality of π was probably the greatest eighteenth-century contribution to the problem of the definite circle quadrature.

In Chapter 8 we saw that Wallis and Gregory had argued that π could not be expressed in algebraic terms, whereas Huygens had maintained that it might turn out to be rational. During the eighteenth century, Huygens' point of view was generally rejected. For example, in his influential textbook *Introductio in Analysin Infinitorum*, Euler declared: "it is obvious that the circumference of this circle (the unit circle) cannot be exactly expressed in rational numbers" (Euler 1748, §126). However, neither he nor anyone else provided an argument for this claim until Lambert published his completed proof in 1768 in a separate publication in French (Lambert 1768).

Lambert's proof built on a continued fraction expansion of $\tan(v)$. First, he proved that if $v = 1/n$ for a natural number n, then the continued fraction would be infinite and having only natural coefficients. It thus results from Euclid's algorithm that its limit, $\tan(v)$, is irrational. By a more complicated argument, Lambert could generalize this result to any rational value of v. Thus, if v is rational, $\tan(v)$ is irrational, or conversely, if $\tan(v)$ is rational, v is irrational. And since $\tan(\pi/4) = 1$, Lambert could conclude that $\pi/4$, and thus π, is irrational.

At the end of his paper of 1768, Lambert even claimed that $\tan(v)$ is transcendental when v is rational: Indeed, having introduced the concept "irrational radical quantity" to mean any root of an algebraic equation with integer coefficients, he continued:

> I say thus that no circular & logarithmic transcendental quantity can be expressed by any irrational radical quantity, which refers to the same unit, & in which there enters no transcendental quantity. The proof of this theorem appears to rest on the fact that the transcendental quantities depend on e^x, where the exponent varies, whereas the radical quantities assume constant exponents.
>
> (Lambert 1768, 72)

From this theorem he concluded that the quadrature of a circle is impossible:

> This theorem being once proven in all its universality, it will follow that since the circumference of the circle cannot be expressed by any radical quantity, nor by any rational quantity, there will be no means to determinate it by some geometric construction. For everything that can be constructed geometrically corresponds to the rational & radical quantities.
>
> (Lambert 1768, 72–3)

These remarks were prophetic. Indeed, the definitive proof of the impossibility of the quadrature of a circle by ruler and compass followed Lambert's plan. However, in order to reach the goal, Lambert's two claims had to be proved rigorously:

1. Constructible quantities are algebraic. As we have seen, this claim was correctly believed to be a rather obvious consequence of the formalism of analytic geometry, but no rigorous proof was offered in the eighteenth century.
2. π is transcendental. This turned out to be more difficult to prove than Lambert seems to have believed. It lasted more than a century before Lindemann delivered the proof.

We shall return to the definitive impossibility proof in Section 13.3.

10
Impossible Equations Made Possible
The Complex Numbers

10.1 The extension of the number system: Wallis's account

This chapter will deal with impossibility statements that are rather simple to prove. The focus is not on the impossibility proofs but on the way mathematicians got around the impossibilities by extending the area of discourse through the introduction of what Hilbert later called ideal elements.

In particular, we shall discuss how the number concept has been extended as an answer to algebraic impossibilities. John Wallis (1616–1703) described this dialectic interplay between impossibility and creativity very accurately in his *Arithmetica Infinitorum* (1656):

> Where some impossibility is arrived at, which indeed must be assumed to be done, but nevertheless cannot actually be done, [the arithmeticians] consider some method of representing what is assumed to be done, though it may not be done in reality.
>
> And this indeed happens in all operations of arithmetic involving resolution, for example, in subtraction: if it is proposed that a larger number must be taken from a smaller, thus 3 from 2 or 2 from 1, since this cannot be shown in reality, there are considered negative numbers, by means of which a supposed subtraction of this kind may be expressed, thus $2 - 3$ or $1 - 2$ or -1.
>
> In division, if it is proposed that a number must be divided by another which is not a divisor, thus 3 by 2, since this cannot be shown in reality, there is invented a method of indicating a supposed division of this kind, in this form $\frac{3}{2}$ or $1\frac{1}{2}$.
>
> In the extraction of roots, if there is proposed a number that is not in its nature truly a power, for example, if there is sought the square root of 12, since that root cannot be expressed as any integer or fractional number, there is invented a method of indicating any supposed root of its kind in this form $\sqrt{12}$ or $2\sqrt{3}$. (Wallis 1656, 161–2)

A History of Mathematical Impossibility. Jesper Lützen, Oxford University Press.
 DOI: 10.1093/oso/9780192867391.003.0010

Wallis was probably the first to describe this dynamic so explicitly. For him the main idea was to find a "representation" or an "expression" for the impossible object: We create a new object by naming it. As we shall see, the question of notation or representation is certainly important but today we would rather emphasize the importance of a suitable environment or mathematical structure in which one can operate with the new object in ways similar to how we operated with the "real" objects whose existence we assumed before the extension. Thus, when we extend the domain of numbers it is important that we can use the normal rules of calculation, e.g., the commutative, associative, and distributive laws on the new objects. For example, as we introduce the negative numbers, it is important to extend the rules of multiplication in such a way that the distributive law continues to hold. That forces us to accept the not so intuitive sign rule: negative times negative gives positive.

In which situations do mathematicians give names (and existence) to impossible objects? According to Wallis it happens when the impossibility "must be assumed to be done." He probably had in mind that the new object should be necessary (or at least useful) for the further argument. It is indeed the case that when mathematicians extend the discourse to a new and larger domain by including previously impossible objects it is because the extension facilitates not just one single problem but a whole family of problems, and in many cases even leads to simplifications of problems which were solvable in the old domain.

As a historical account, Wallis's description of the gradual extension of the domain of number is highly rationalized. In particular it is very formal or nominalist. In fact, the acceptance of new types of numbers has usually been accompanied by some sort of intuitive interpretation.

Negative numbers were introduced first in China about year 0. They were represented by black calculating rods and the positive numbers were represented by red rods. Negative numbers were used in the so-called fang cheng method of solving a system of linear equations (similar to Gaussian elimination). In this method, it is important to allow negative numbers in the intermediate steps, even when the result turns out to be positive. The Chinese also had an interpretation of the negative numbers: If a positive number n means that one buys n cows, the negative number $(-n)$ means that one sells n cows. Today we usually illustrate negative numbers geometrically as numbers lying on a number line in the opposite direction from the positive numbers. This representation goes back to Wallis' *Arithmetica Infinitorum*.

Negative numbers also appeared in medieval India. Here they arose out of the rules for calculating with expressions of the form $(a\text{-}b)$. For example, in order to explain how to multiply two numbers of this kind with each other, Brahmagupta in his *Brāhmasphuṭasiddhānta* (AD 628) explicitly formulated the sign rules, e.g., the product of two negatives is positive. Starting from such rules, the negative numbers (and zero) gradually took on a life of their own.

In medieval Europe, negative numbers appear in a famous problem in Leonardo Fibonacci's *Liber Abaci* (1202). The problem deals with five men, each having an unknown amount of money that one must determine from five equations. Leonardo shows that the system of equations cannot be solved unless one assumes that one of the men has a negative amount of money, interpreted as a debt. In the Renaissance the sign rules were formulated by abacists who practiced the art of algebra using Hindu-Arabic numerals. As in India, they led to a gradual acceptance of negative numbers. For example, Nicolas Chuquet (*c.*1450–*c.*1490) sometimes accepted negative solutions of equations but at other times he rejected them as impossible (Katz 2009, 392). In fact, it was well into the seventeenth century before negative numbers were accepted by most mathematicians, and even as late as the nineteenth century, leading mathematicians engaged in a debate about the existence of negative numbers (Schubring 2005).

Superficial as it may be, this historical account reveals that the development of negative numbers only follow Wallis' pattern to a certain degree. Fibonacci and Chuquet were faced with problems that could only be solved by introducing negative numbers. So they followed Wallis' model, except for Wallis' insistence on the importance of notation. The rest of the story does not follow Wallis' pattern at all. In fact, in the Indian and the Renaissance story it is the algebraic operations that are developed first; the generalization of the objects follow as a consequence.

Historically, Wallis' account of the introduction of fractions and surds is even further from the historical development. As we saw, the discovery of incommensurable quantities in antiquity did not make the Greeks introduce surds. Instead, it made them distinguish between numbers, on the one hand, and geometric quantities that they studied without assigning numbers to them, on the other hand. And although applied mathematicians used fractions, the pure geometers developed a theory of ratios that could accommodate geometric magnitudes as well as numbers. In the late Middle Ages, mathematicians began to explore the analogy between ratios of magnitudes and fractions, but it was only around 1700 that fractions replaced ratios, and surds and irrational numbers were used freely.

There is another extension of the number concept that follows Wallis' pattern more closely, namely the introduction of the complex numbers. We shall analyze this story in some detail, also because it connects nicely to our earlier analysis of diorisms, in particular the elliptic application of areas.

10.2 Cardano's sophisticated and useless numbers

Already in the Arab medieval sources it is explicitly noticed that quadratic equations may not possess a solution, even when one accepts negative solutions. In al-Khwārizmi's famous systematic account of solutions of quadratic equations (about AD 800) he mentioned that an equation of the form

$$x^2 + c = bx \tag{10.1}$$

has a solution that can be found by an algorithm that we can represent by the formula

$$x = \frac{b}{2} \pm \sqrt{\left(\frac{b}{2}\right)^2 - c}. \tag{10.2}$$

When $c > \left(\frac{b}{2}\right)^2$ so that the number $\left(\frac{b}{2}\right)^2 - c$ under the square root sign is negative, al-Khwārizmi declared that the problem is impossible. His contemporary Ibn Turk more pointedly mentioned that "there is the logical necessity of impossibility in this type of equation when the numerical quantity ... is greater than (the square of) half the number of roots" (Katz 2009, 274). This impossibility was circumvented by the introduction of complex numbers.

The first recorded introduction of complex numbers is found in *Ars Magna* published in 1545 by Gerolamo Cardano (Figure 10.1). Complex numbers appear twice. In the thirty-eighth problem Cardano briefly remarked:

> Note that $\sqrt{9}$ is either +3 or -3, for a plus [times a plus] and a minus times a minus yields a plus. Therefore $\sqrt{-9}$ is neither +3 nor -3 but is some recondite third sort of thing.
>
> (Cardano 1968, 219)

Cardano here hints at the general proof of the impossibility of determining a (real) square root of a negative number or equivalently a solution of the equation $x^2 = -a$ when a is a positive number: x cannot be positive, negative, or zero. For if x is positive or negative x^2 is positive and if x is 0, $x^2 = 0$. So x^2 cannot be negative.

Figure 10.1 Gerolamo Cardano (1501–76)
Reproduced by Jim Høyer DMS

In Chapter 37 of the *Ars Magna* entitled "On the Rule for Postulating a Negative," he discussed a more interesting problem leading to complex solutions. Having discussed how to handle negative solutions to equations he turned to "a second species of negative assumption [involving] the square root of a negative":

> I will give an example: If it should be said, Divide 10 into two parts the product of which is 30 or 40, it is clear that this case is impossible. Nevertheless, we will work thus: We divide 10 into two equal parts, making each 5. These we square, making 25. Subtract 40, if you will, from the product thus produced, as I showed you in the chapter on operations in the sixth book, leaving a remainder of −15, the square root of which added to or subtracted from 5 gives parts the product of which is 40. These will be $5+\sqrt{-15}$ and $5-\sqrt{-15}$.
> (Cardano 1968, 219–20)

Here Cardano tries to solve a system of equations of the form:

$$x + y = b, \tag{10.3}$$

$$x \cdot y = c. \tag{10.4}$$

This system of equations is equivalent to the quadratic equation (10.1). We have also met systems of this type in the Babylonian sources and in a somewhat different form in the elliptic application of areas in Euclid. The latter had (in a more general setting) proved that this system has solutions if and only if the diorism $c \leq \left(\frac{b}{2}\right)^2$ is satisfied. In the case investigated by Cardano, $b = 10$ so there exists no solution unless $c \leq \left(\frac{10}{2}\right)^2 = 25$ However, Cardano has chosen $c = 40$ which is definitely greater than 25, so as he points out, it is clear that this case is impossible.

Still, he tries to follow the solution algorithm already known by the Babylonian scribes and the Arab mathematicians and he arrives at the solution $x = 5 + \sqrt{-15}$ and $y = 5 - \sqrt{-15}$.

What made it possible to find a solution where the Babylonians and the Arabs had failed? The crucial novelty that Cardano had at his disposal was a new rudimentary algebraic notation, in particular a sign for negative numbers and for square root. His square root sign was a capital R with an attached x signifying "radix." It seems to have been the origin of our square root sign, which is an *r* followed by a horizontal line. The Babylonians and the Arabs would have been blocked in their use of the algorithm at the point where they had to subtract 40 from 25. They had no way to write the result. And even if, like the Indians, they had had a notation for negative numbers, they would have been blocked in the next step where they should take the square root of the difference −15. Cardano, on the other hand, could just write $\sqrt{-15}$.

The importance of the notation is beautifully illustrated by Cardano's proof of his solution of the problem. Following the tradition even among algebraists, Cardano tried to give a geometric proof following closely Euclid's construction of the elliptic application of areas (Figure 6.2). So he set out a line segment *AB* equal to 10 and divided it in two equal parts in *C*. Then he constructed the square on *AC*, as prescribed by Euclid (Figure 6.2). Now, according to Euclid, Cardano should place a square (the grey square in Figure 6.2) of area 25–40=−15 in the corner of the square on *AC*. At this point of the demonstration, Cardano simply gave up the geometric approach. Indeed how should he have represented a square with a negative area? Instead he continued his argument in a purely algebraic way representing the side of the square by $\sqrt{-15}$.

It is often maintained that a mathematical result can usually be presented in many different forms having the same content. In particular, in the debate about "geometric algebra," Zeuthen and Van der Waerden and others have pointed out that the same algebraic content can be expressed in geometric terms, as the Greeks did or in arithmetic-algebraic terms as the Babylonians (perhaps) and the Arabs did. Sabetai Unguru, on the other hand, does not think it makes any sense to speak about a mathematical content independent of its form (Unguru 1975–6). While I think Unguru goes too far in his criticism, I think that Cardano's introduction of the complex numbers is a clear example where the geometric and the algebraic formulations of a problem are not equivalent. The Euclidean geometric style offers no obvious way to represent a square with a negative area, whereas the algebraic notation allows Cardano to write a symbol $\sqrt{-15}$. This is exactly what Wallis later wrote that one should do when one meets an impossibility: Invent a notation for the impossible object. What Wallis had not stressed, but what Cardano then went on to do, is to extend the algebraic methods of operation to the new symbols. Indeed, he showed that the sum of the two solutions he had found is 10 and their product is 40.

Cardano does not seem to have appreciated the importance of his new symbols. At least he concluded his solution with the words: "So progresses arithmetic subtly the end of which, as is said, is as refined as it is useless" (Cardano 1968, 220).

10.3 The unreasonable usefulness of the complex numbers

If the complex numbers had been useless as Cardano thought, the problem in Section 10.2 and Cardano's "solution" would not have made it into this book or other books on the history of mathematics. Indeed, even if Cardano could write arithmetic expressions that formally satisfy the problem, it does not help anybody who wants to divide a line of length 10 into two segments spanning a rectangle with area 40. To put it more generally: As already remarked by al-Khwārizmi, when the solution formula for the quadratic equation contains a square root of a negative number (the discriminant), the equation has no (real) solutions. This could easily have been the end of the story. But the solution of the cubic equation changed the situation.

Cardano's *Ars Magna* is primarily famous for its algorithm for calculating the solutions of the cubic and biquadratic equations (polynomial equations of degree 3 and 4). The solution of the cubic equation that the Arabs had sought in vain was the first great mathematical accomplishment of European scholars after a long medieval period in the shadow of the flourishing Islamic culture. The first case was solved by Scipione del Ferro (1465–1526) around 1510, the next by Niccolò Tartaglia (1500–57) from whom Cardano learned the solution. Violating his promise to Tartaglia, Cardano published the algorithm for all types of cubic equations in his *Ars Magna*, and supplied them with proofs of his own. In the same book, he also published his student Ferrari's solution of the biquadratic equation.

In order to understand the importance of the complex numbers for the solution of cubic equations we shall go into a few details. In modern notation, a general cubic equation can be written as

$$ax^3 + bx^2 + cx + d = 0,$$

where a, b, c, d are real numbers. Cardano had no notation for a general coefficient, so he always solved equations with specific numerical values of the coefficients a, b, c, d. However, his presentation of the algorithm was phrased in such a way that the general rule became clear to the reader. Moreover, even though Cardano did allow negative solutions, he always chose the coefficients a, b, c, d to be positive. In case where we would have a negative coefficient, he placed the corresponding term on the right side of the equation. Thus, he had to distinguish between many different cases of cubic equations just as al-Khwārizmi had distinguished between several types of quadratic equations.

Even before Cardano, Italian algebraists knew how to get rid of the x^2 term of a cubic equation by a simple linear substitution. This reduced the number of interesting cases to three. Let us consider an equation of the form

$$x^3 = cx + d.$$

In this case Cardano's solution algorithm corresponds to the modern formula for the solution, called Cardano's formula:

$$x = \sqrt[3]{\frac{d}{2} + \sqrt{\left(\frac{d}{2}\right)^2 - \left(\frac{c}{3}\right)^3}} + \sqrt[3]{\frac{d}{2} - \sqrt{\left(\frac{d}{2}\right)^2 - \left(\frac{c}{3}\right)^3}}.$$

Already Cardano discovered that this formula is problematic when the value under the square root sign is negative, i.e., when

$$\left(\frac{c}{3}\right)^3 > \left(\frac{d}{2}\right)^2.$$

This is the so-called irreducible case. If there had been no real solutions in this case, as in the similar situation for a quadratic equation, there would have been no problem. However, it turns out that in the irreducible case, there are, in fact, three real solutions. Thus, Cardano's formula apparently fails in the irreducible case.

Except it does not. Fifteen years after the appearance of Cardano's *Ars Magna*, Rafael Bombelli (1526–72) wrote a new *Algebra* in which he argued that if we calculate correctly with the square root of negative numbers, Cardano's formula is valid also in the irreducible case. For example, he considered the equation

$$x^3 = 15x + 4.$$

It is easy to verify that 4 is a solution to the equation, and yet Cardano's formula gives

$$x = \sqrt[3]{2 + \sqrt{-121}} + \sqrt[3]{2 - \sqrt{-121}}.$$

Bombelli wanted to show that this expression is indeed equal to 4. Postulating that $\sqrt[3]{2 + \sqrt{-121}} = a + \sqrt{-b}$ and that $\sqrt[3]{2 - \sqrt{-121}} = a - \sqrt{-b}$ he argued (with a bit of luck) that $a = 2$ and $b = 1$, so that $x = \left(2 + \sqrt{-1}\right) + \left(2 - \sqrt{-1}\right) = 4$. Thus, even if the solution is real, the calculation using Cardano's formula takes us through the type of numbers that were later called complex numbers.

Examples like this gradually convinced mathematicians that complex numbers were not useless sophistries as Cardano had thought, but a very useful new type of number. At first, they showed their great usefulness in the theory of equations. Indeed, it turned out that any polynomial equation of degree n has exactly n complex solutions if they are counted suitably with multiplicity. That is the famous fundamental theorem of algebra. The first mathematicians who formulated a variant of this theorem were Albert Girard (1595–1632) and Descartes in 1629 and 1637, respectively. They both claimed that one can imagine that an nth degree equation has n solutions. They did not state that these roots were all of the form $a + \sqrt{-b}$ or $a + b\sqrt{-1}$, and indeed Leibniz argued that there could be roots of a more complicated kind. For example, the equation

$$x^4 + 1 = 0$$

has solutions of the kind $\pm\sqrt{\pm\sqrt{-1}}$ which Leibniz did not think was of the form $a + b\sqrt{-1}$. Soon, however, Nicolas Bernoulli could show that $\sqrt{\sqrt{-1}} = \left(1 + \sqrt{-1}\right) \cdot \frac{1}{\sqrt{2}}$, so it is a complex number after all. In the eighteenth century some of the greatest mathematicians like d'Alembert and Euler tried to prove that all the n imagined roots in an nth degree equation are complex numbers. However, the first widely accepted proofs were given by Carl

Friedrich Gauss (1777–1855) in 1799, 1814/15, 1816, and 1850. In the last of his four proofs, Gauss even proved the fundamental theorem in the case where the polynomial has complex coefficients.

Already in the eighteenth century Euler showed how complex numbers could be used in analysis. In particular he discovered a beautiful and completely surprising connection between the exponential function and trigonometric functions: $e^{iz} = \cos z + i \sin z$. This is now called Euler's formula. In the nineteenth century Augustin Louis Cauchy (1789–1857), Bernhard Riemann (1826–66), and Karl Weierstrass (1815–97) continued Euler's work, developing a whole new area of complex function theory. About this same period complex numbers found their way into physics, first in optics and then in the theory of alternate currents and quantum mechanics.

It is remarkable that 250 years passed before complex numbers got an interpretation. From the start of their life in Cardano's book, complex numbers were nominally defined by a notation (as Wallis would have it) and rules describing how to operate with them, but it was only around 1800 that they were given a geometric interpretation as numbers in a complex number plane (Caspar Wessel, Argand, Gauss). This sets their history apart from the history of other number concepts.

The invention of complex numbers is probably the most successful example of the method of ideal elements as a way around an impossible problem. By introducing the notation $\sqrt{-1}$ (which Euler and Gauss called i) as the solution of a particular quadratic equation without a solution ($x^2 + 1 = 0$), mathematicians have created an entirely new world of numbers that has enriched much of mathematics and physics. Not only did nth degree equations now always get n solutions (which is very useful in many connections) but there appeared a new connection between the exponential function and trigonometric functions. Moreover, physicists obtained a language that was useful in the theory of electricity and almost necessary for the formulation of quantum mechanics. For this (and other) reasons the physicist Eugene Wigner (1960) has spoken of the unreasonable effectiveness of mathematics. Complex numbers are indeed the most striking example of a mathematical theory that was created for a purely mathematical purpose and later turned out to be extremely important in physical applications.

10.4 A digression about infinitesimals

It is interesting to compare the development of complex numbers with the development of infinitesimals. Infinitesimals or infinitely small quantities were introduced in the first half of the seventeenth century as a tool for finding tangents, areas, and similar geometric objects related to curves. They became key notions in the differential and integral calculus invented by Leibniz in the 1670s and in a variant form by Newton a bit earlier. Leibniz denoted an infinitesimal increase in a variable x by dx and introduced rules for calculating with such differentials. Using these rules, he could almost automatically derive results that would be difficult to obtain with earlier methods. Infinitesimals were often defined as quantities that were smaller than any finite quantity. But do such quantities exist or are they impossible? This question was hotly debated in various fora such as the French Academy of Sciences. Leibniz admitted that infinitely small line segments do not exist in nature, nor do they exist in the same sense as finite line segments. But he maintained that they could be introduced as fictitious or imaginary quantities with which one can calculate. He also maintained that if one follows the rules of the differential calculus one will have a speedy and secure way for deducing true theorems. Therefore, they were an invaluable tool for the advancement of mathematics. (see Bos 1974).

Leibniz often likened infinitesimals to imaginary or complex numbers. Complex numbers do not exist in the same way as other numbers, but they can simplify algebra, and results found using them are always correct. From a modern perspective infinitesimals and complex numbers are both examples of what Hilbert called ideal elements. However, their later reception followed different patterns. The acceptance of complex numbers by the mathematical community had a continuous increasing trend until the middle of the nineteenth century where they were generally accepted by all professional mathematicians. The acceptance of infinitesimals followed a much more bumpy road. At first, they were criticized for being incomprehensible and inconsistent. How could infinitesimals like dx sometime be considered as being non-zero, so that one can divide by them, and then later in the same calculation they are set equal to zero. As the philosopher and Bishop George Berkeley (1685–1753) polemically put it in 1734:

> And what are these ... evanescent increments? They are neither finite quantities, nor quantities infinitely small, nor yet nothing. May we not call them the ghosts of departed quantities.
>
> (Berkeley 1734, in Smith 1959, 627)

During the eighteenth century the critical voices were almost silenced by Euler and his contemporaries' successful application of differentials in the mathematical description of natural phenomena. However, toward the end of the century the requirement of more rigor was raised by Joseph-Louis Lagrange (1736–1813). In his approach to the calculus of functions, the notion of the derivative replaced the differentials as the fundamental concept. In Cauchy's influential textbooks written for students of the French *École Polytechnique* in the 1920s, infinitesimals continued to appear but they changed their meaning from being actually infinitely small to being variables converging to zero. Finally, with Weierstrass's trendsetting lectures around 1870 in Berlin, the idea of infinitesimally small quantities was altogether removed from the calculus. It was replaced by the $\varepsilon - \delta$ definitions and proofs we still use in the rigorous modern approach to analysis.

In applications of mathematics infinitesimals continued to be used as intuitive but inconsistent quantities until the middle of the twentieth century where they were resurrected as genuine mathematical objects. In 1966 Abraham Robinson (1918–74) showed that it is possible to enrich the real numbers by infinitesimals in a consistent way.[1] In the resulting universe of non-standard analysis, one can apply the usual rules of operation as with real numbers except the Archimedean property: Given a finite real number and an infinitesimal there does not exist a natural number whose product with the infinitesimal exceeds the given finite real number. Robinson argued that his non-standard analysis vindicated Leibniz', Euler's, and other earlier mathematicians' calculations using infinitesimals. This claim has been challenged, in particular because Robinson's construction of his new non-standard universe used modern methods that were far out of the reach of the earlier mathematicians. Moreover, non-standard analysis makes distinctions that earlier mathematicians did not make. For example, Euler would write $dx = 0$ where a modern non-standard analyst would distinguish between equality and "being infinitely close to."

Be that as it may. The creation of non-standard analysis is another example of how the introduction of ideal elements can circumvent otherwise impossible problems or concepts, here infinitesimals.

[1] In 1958 Curt Schmieden and Detlef Laugwitz published a similar, but not quite as satisfactory infinitesimal calculus.

11

Euler and the Bridges of Königsberg

> The problem, which I am told is widely known, is as follows: In Königsberg in Prussia, there is an island *A* called the Kneiphof; the river which surrounds it is divided into two branches, as can be seen in Figure 11.1, and these branches are crossed by seven bridges, *a*, *b*, *c*, *d*, *e*, *f*, and *g*. Concerning these bridges it was asked whether anyone could arrange a route in such a way that he would cross each bridge once and only once. I was told that some people asserted that this was impossible, while others were in doubt; but nobody would actually assert that it could be done. From this I have formulated the general problem: whatever be the arrangement and division of the river into branches, and however many bridges there be, can one find out whether it is possible to cross each bridge exactly once?
>
> (Euler 1736 in Biggs et al. 1976, 3)

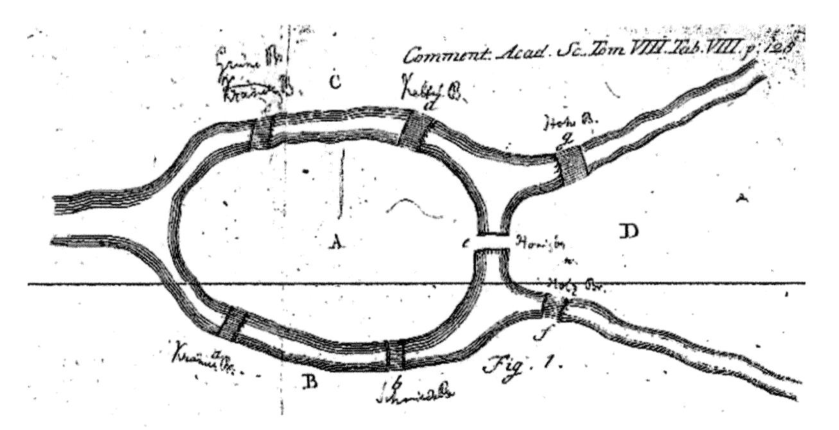

Figure 11.1 Euler's illustration of the seven bridges of Königsberg
Reproduced by Jim Høyer from Euler's Opera Omnia Series 1, Volume 7

In this way, Leonhard Euler (Figure 11.2) presented the problem of the seven bridges of Königsberg (now Kaliningrad) in a 1736 paper with the title "The Solution of a Problem Relating to the Geometry of Position." By then the Swiss-born mathematician worked at the Scientific Academy in St. Petersburg.

A History of Mathematical Impossibility. Jesper Lützen, Oxford University Press.
 DOI: 10.1093/oso/9780192867391.003.0011

He became the leading mathematician of the eighteenth century, publishing profusely in all branches of pure and applied mathematics. Being a master of calculations, Euler at first did not recognize the problem of the bridges of Königsberg as a mathematical problem at all. When his friend Carl Gottlieb Ehler in a letter of 1736 asked Euler about a solution to the problem Euler answered:

> Thus you see, most noble Sir, how this type of solution bears little relationship to mathematics, and I do not understand why you expect a mathematician to produce it, rather than anyone else, for the solution is based on reason alone, and its discovery does not depend on any mathematical principle. Because of this, I do not know why even questions which bear so little relationship to mathematics are solved more quickly by mathematicians than by others. In the meantime, most noble Sir, you have assigned this question to the geometry of position, but I am ignorant as to what this new discipline involves, and as to which types of problem Leibniz and Wolff expected to see expressed in this way.
>
> (Sachs et al. 1988, 136)

It was Ehler who had mentioned that the problem was an outstanding example of a new area of mathematics, called calculus situs, which Leibniz had described as "yet another kind of analysis, geometric or linear, which deals directly with position, as algebra deals with magnitude" (Sachs et al. 1988, 134). Despite his initial skepticism, Euler became convinced of Ehler's point of view, and in his published paper, he presented the Königsberg problem as an illustration of what Leibniz could have had in mind when he introduced the new vaguely described branch of mathematics:

> In addition to that branch of geometry which is concerned with magnitudes, and which has always received the greatest attention, there is another branch, previously almost unknown, which Leibniz first mentioned, calling it the geometry of position. This branch is concerned only with the determination of position and its properties; it does not involve measurements, nor calculations made with them. It has not yet been satisfactorily determined what kind of problems are relevant to this geometry of position, or what methods should be used in solving them. Hence, when a problem was recently mentioned, which seemed geometrical but was so constructed that it did not require the measurement of distances, nor did calculation help at all, I had

> no doubt that it was concerned with the geometry of position—especially as its solution involved only position, and no calculation was of any use. I have therefore decided to give here the method which I have found for solving this kind of problem, as an example of the geometry of position.
>
> (Euler 1736 in Biggs et al. 1976, 3)

In order to solve the problem, Euler denoted the different land areas by capital letters *A*, *B*, *C*, ... and the bridges between them by small letters *a*, *b*, *c*, ... A route beginning in *A* crossing a bridge into *B*, continuing over a new bridge into *D* and further via a new bridge into *C*, etc., is then denoted by a sequence of capital letters: *ABDC* ... . At first Euler did not specify which of the bridges is crossed if there are more than one bridge between two land areas. The only important thing we need to know is that each bridge is crossed exactly once.

Figure 11.2 Leonhard Euler (1707–83)
Reproduced by Jim Høyer from Euler's Opera Omnia Series 2, Volume 1

If there exists a tour crossing each of the bridges exactly once (a so-called Euler tour) it is clear that the number of capital letters in the sequence of land areas must contain one letter more than the total number of bridges. Thus, in the problem of the Königsberg bridges the tour must be denoted by a sequence of 8 capital letters.

On the other hand, the number of times a capital letter, say *A*, occurs in the sequence denoting the tour is determined by the number of bridges leading to that land area. If there is an odd number n_A of bridges leading to *A*, then *A* must occur $\frac{n_A+1}{2}$ times in the sequence. For example, if there are three bridges leading to *A* and the tour starts in *A*, it must come back to *A* precisely once.

Indeed, we will start by leaving A by one bridge, and when we come back, we will enter A by one bridge and then leave A by another. If, on the other hand, the tour begins elsewhere than A, the first occurrence of A will need one entering bridge and one leaving bridge, and then we will need to come back to A once more to cross the third bridge.

Similarly, if the number n_A of bridges leading to A is even, then A must occur $\frac{n_A}{2}$ times in the sequence if the tour does not start in A, and $\frac{n_A}{2} + 1$ times if it starts in A.

In the Königsberg case, $n_A = 5, n_B = 3, n_C = 3, n_D = 3$, so Euler concluded that A must occur three times, and the other three letters twice. That gives a total of nine letters. But from the first way of counting he knew that the tour must contain eight letters. Thus, the desired tour does not exist.

In general Euler suggested the following method: For each land area X with an odd number n_X of bridges leading to it, write down the number $\frac{n_X+1}{2}$, and for each land area Y with an even number of bridges write down the number $\frac{n_Y}{2}$. Add the numbers to obtain a sum S. If an Euler tour starts in an area with an odd number of bridges this sum must be equal to the total number of bridges plus one. If the tour starts in one of the areas with an even number of bridges, the sum S must be equal to the number of bridges. If S in not equal to the number of bridges or to the number of bridges plus one, an Euler tour does not exist.

In the last part of the argument, I have repaired Euler's logic. In fact, he wrote:

> and if this sum [S] is one less than, or equal to, the number written above, which is the number of bridges plus one, I conclude that the required journey is possible. It must be remembered that when the sum is one less than the number written above, then the journey must begin from one of the areas marked with an asterisk [those with an even number of bridges], and it must begin from an unmarked one if the sum is equal.
>
> (Euler 1736 in Biggs et al. 1976, 6)

Thus, Euler concluded that his condition is sufficient for the existence of an Euler tour. However, his argument only shows that the condition is necessary. In other words, his proof of the counting method can be used to establish that an Euler tour does not exist, as in the Königsberg case. In fact, it is also a sufficient condition, but Euler's argument does not prove it.

Euler further simplified the condition for the existence of an Euler tour.

> First, I observe that the numbers of bridges written next to the letters A, B, C, etc [i.e. the numbers n_A, n_B, n_C, etc] together add up to twice the total number of bridges. The reason for this is that, in the calculation where every bridge leading to a given area is counted, each bridge is counted twice, once for each of the two areas which it joins.
>
> (Euler 1736 in Biggs et al. 1976, 7)

From this fact, Euler concluded that if one adds the number of bridges leading to all the areas the result is an even number and therefore there must be an even number of areas with an odd number of bridges leading to it. Recall that we formed the sum S by adding the numbers $\frac{n_X+1}{2}$ for areas X with an odd number of bridges, and $\frac{n_Y}{2}$ for areas Y with an even number of bridges. This sum total will be ½ times the total number of bridges leading to an area (i.e., the total number of bridges) plus ½ times the number of areas with an odd number of bridges (which is an even number). Thus, if there are no areas with an odd number of bridges, the sum is equal to the number of bridges and if there are two areas with an odd number of bridges the sum is the sum of the bridges plus one. In these cases, Euler's condition is satisfied. If there are more than two areas with an odd number of bridges, the condition is not satisfied. Euler concluded:

> So whatever arrangement may be proposed, one can easily determine whether or not a journey can be made, crossing each bridge once, by the following rules:
> If there are more than two areas to which an odd number of bridges lead, then such a journey is impossible.
> If, however, the number of bridges is odd for exactly two areas, then the journey is possible, if it starts in either of these areas.
> If, finally, there are no areas to which an odd number of bridges leads, then the required journey can be accomplished starting from any area.
>
> (Euler 1736 in Biggs et al. 1976, 8)

As remarked earlier, Euler's conclusions are correct but his proof only guarantees the first impossibility part.

At the end of the nineteenth century questions like the Königsberg problem was formulated in terms of what James Joseph Sylvester (1814–97) called a graph. It is a collection of so-called vertices (think of Euler's land areas) illustrated by points, and so-called edges joining two vertices (think of Euler's bridges) illustrated by curves. In this graphical language, the layout of the bridges in Königsberg can be illustrated as in Figure 11.3.

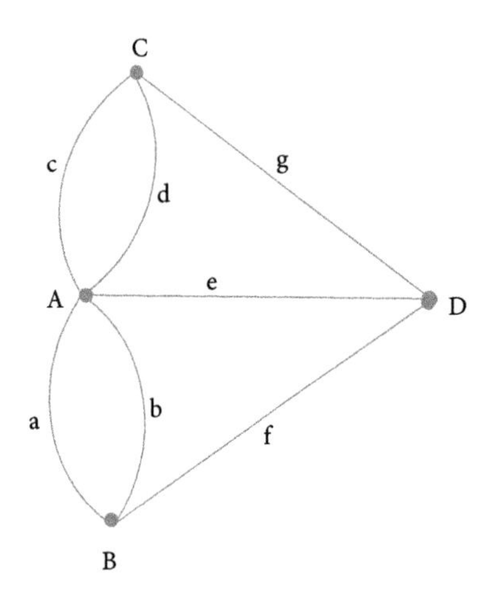

Figure 11.3 Graph illustrating the bridges of Königsberg

It is often claimed that this way of illustrating the situation in Königsberg is due to Euler but it is much later. The number of edges connected with a vertex is called the valency of the vertex (a notation due to Sylvester, who used graphs to illustrate molecules). The sum of the valences of all the vertexes is called the total valence of the graph. As noted by Euler, this is twice the number of edges. Also the proof of the necessity of Euler's condition for the existence of an Euler tour was later simplified. The modern standard proof by Carl Hierholzer (1840–71) was published in 1873, two years after his death (see Biggs et al. 1976, 11). Here it is:

a. If an Euler tour in a graph (i.e., a tour using all the edges exactly once) *starts and ends at different vertices*, the valency of all vertices except the start vertex and end vertex will be an even number. Indeed, the vertices come in pairs: every time the tour reaches one of these vertices, it will arrive by one edge and leave by another. This hold true also for the start vertex and end vertex, except for the beginning edge and the end edge. Thus, the start vertex and the end vertex will have odd valences. In this case, therefore, there are exactly two vertices with odd valences.
b. If, on the other hand, the tour begins and ends in the same vertex (a closed Euler tour), all the vertices have even valence. Indeed, in this case the beginning edge and the end edge will form a pair for the vertex where the tour begins and ends.

Thus, we can conclude as Euler did, that if a graph has more than two vertices with odd valence, it is impossible to find an Euler tour in it.

Hierholzer also proved the converse existential theorem formulated, but not proved, by Euler to the effect that if there are two or no vertices with odd valency, there exists an Euler tour in the graph. This proof is more complicated, and the result is more surprising.

Euler's paper on the bridges of Königsberg is very atypical for the time and for Euler in particular. First, as emphasized by Euler himself, the paper was the first to deal with a problem in the new branch of geometry called geometria situs by Leibniz. This type of geometry, where distances were not important, continued to develop as a marginal part of mathematics until the end of the nineteenth century where it gave rise to two new very active disciplines: graph theory and topology.

Second, the focus on a question of existence or impossibility is hardly found elsewhere in Euler's remaining enormous production. Euler was a constructive mathematician. For example, in connection with differential equations he always wanted to find an explicit formula for the solution of a given equation. The more fundamental question of the existence of a solution was not raised until the nineteenth century. In the problem of the Königsberg bridges, on the other hand, Euler only presented a condition for the solvability or impossibility of the solution, a diorism, one may say. The question of how to construct an Euler tour if it exists, was only mentioned in a few lines at the end of Euler's paper. He simply asked his reader to remove all pairs of bridges between the same two areas, and then claimed that it would be easy to find an Euler tour in the remaining part of the figure (graph). This is not obvious but in a sense true. Indeed, in 1883 Fleury showed that if one begins the tour in a vertex with odd valency (or at an arbitrary vertex, if all vertices have even valency) and then proceed in an almost arbitrary way along unused edges, one will end up with an Euler tour. The only thing one cannot do is to use an edge that makes the graph disconnected if one can avoid it. Another algorithm is implicitly contained in Hierholzer's proof.

12
The Insolvability of the Quintic by Radicals

12.1 Early results

The solution of polynomial equations has played a central role in many cultures. We have seen that the Babylonians were able to solve quadratic equations already around 2000 BC and that Cardano in 1545 published the solutions of the cubic and the biquadratic equations in his *Ars Magna*. The solutions were formulated as algorithms that with later seventeenth-century notation could be written as algebraic formulas expressed in terms of the coefficients of the equation using the algebraic operations $+$, $-$, $\cdot$, $:$, $\sqrt[n]{\ }$. Since nth root signs are also called radicals, one speaks of a solution by radicals. For example, Cardano's formula

$$x = \sqrt[3]{\frac{d}{2} + \sqrt{\left(\frac{d}{2}\right)^2 - \left(\frac{c}{3}\right)^3}} + \sqrt[3]{\frac{d}{2} - \sqrt{\left(\frac{d}{2}\right)^2 - \left(\frac{c}{3}\right)^3}}$$

for the solution of the cubic equation

$$x^3 = cx + d$$

is a solution by radicals, sometimes also called an algebraic solution.

Cardano's success gave hopes of finding similar solutions by radicals of equations of degree 5 (the quintic) and higher, but that turned out to be more difficult than expected. During the first one and a half centuries after Cardano, little progress was made but during the eighteenth century, Euler, Etienne Bézout (1730–83), Edward Waring (1734–89), Alexandre-Théophile Vandermonde (1735–96), and Joseph-Louis Lagrange (1736–1813) published important investigations. The last three published their contributions in the year 1770, the most influential being the book-length paper by Lagrange.[1]

Although they did not lead to a solution, these eighteenth-century works introduced ideas that were crucial in the final attack on the problem. One

[1] Stedall (2011) gives a detailed study of the development of the theory of equations in the period between Cardano and Lagrange.

A History of Mathematical Impossibility. Jesper Lützen, Oxford University Press.
 DOI: 10.1093/oso/9780192867391.003.0012

of these ideas was the concept of a resolvent. It goes back to Ferraro's solution of the biquadratic equation. Ferraro discovered that he could solve the biquadratic equation by first solving a cubic equation (the resolvent) and then solve a quadratic equation whose coefficients were rational in the solution of the resolvent. This led to the search for a resolvent equation for the quintic, i.e., an equation whose solution would reduce the degree to 4 or less. Euler and Bézout tried out this strategy, but came up with resolvents of degree higher than 5.

Another important idea was to consider functions of the roots of the equation that have few values under permutations of the roots. This idea can be found in rudimentary form in works by Euler and Waring but was developed much further by Lagrange (Figure 12.1). It is based on the fact, first emphasized by Girard (1629), that the coefficients in a polynomial equation can be expressed in a simple way in terms of the roots. Indeed, consider the polynomial equation of degree n with the roots $r_1, r_2, \ldots, r_n$:

$$P(x) = x^n + a_{n-1}x^{n-1} + a_{n-2}x^{n-2} + \cdots + a_1 x + a_0 = 0.$$

If $r_1, r_2, \ldots, r_n$ denote the roots written as many times as their multiplicity, the polynomial can be factored into first-degree factors[2] as

$$P(x) = (x - r_1)(x - r_2)\cdots(x - r_n).$$

Multiplying out the right-hand side leads to Girard's result:

$$a_{n-1} = -(r_1 + r_2 + \cdots + r_n)$$

$$a_{n-2} = r_1r_2 + r_1r_3 + \cdots r_1r_n + r_2r_3 + \cdots r_{n-1}r_n$$

$$a_k = (-1)^{n-k}\,(\textit{sum of all different products of } (n-k) \textit{ different roots})$$

$$a_1 = (-1)^{n-1}(r_1r_2\cdots r_{n-1} + \cdots + r_2r_3\cdots r_n)$$

$$a_0 = (-1)^n r_1r_2\cdots r_n.$$

The right-hand sides are the so-called elementary symmetric polynomials in $r_1, r_2, \ldots, r_n$. They are, in fact, symmetric in the sense that they do not

[2] This was known by Harriot and Descartes and probably also by Girard (Stedall 2011, 39; Katz 2009, 473). It is necessary to include all complex roots. That is why Girard needed his version of the fundamental theorem of algebra.

change if the roots $r_1, r_2, \ldots, r_n$ are permuted. Moreover, any symmetric rational function in the roots can be written as a rational function in the elementary symmetric polynomials, i.e., in the coefficients. This fundamental property was more or less established by Newton. Since the coefficients of the polynomial are considered known, all symmetric functions of the roots are known quantities.

Figure 12.1 Joseph Louis Lagrange (1736–1813)
Reproduced by Jim Høyer DMS

Moreover, if a polynomial in the roots has k values under permutations of the roots, this polynomial is a solution of a kth degree equation (a resolvent equation) with known coefficients, i.e., coefficients that are rational in the coefficients of the given equation. According to Lagrange, this reveals the "metaphysics" or the "true principles" of equation solving. He came to this conclusion by analyzing the known methods for solving equations of degree 2, 3, and 4, concluding that in every case the method worked because one could find a rational function of the roots, having few values when the roots are permuted among themselves.

In the case of the quintic, Lagrange suggested that one starts by finding a rational function of the roots having few (say k) values under permutations of the roots. This function is then a solution of a kth degree equation with known coefficients. If we know how to solve this resolvent equation, for example if $k \leq 4$, we can assume its solutions, i.e., the rational function, are known. One must then seek another rational function having few (say l) values under those permutations of the roots that fix the first rational function. This second function will then be a root of a polynomial equation of degree l and with coefficients that are rational among the original coefficients and the first (now known) function. After a number of such steps, one may be lucky that one of the roots of the original equation has few values under the permutations that fix the last rational function. It will then be a solution of a low-degree resolvent.

So if all the mentioned rational functions have less than five values under the said permutations we can solve all the resolvent equations by radicals. The result will be a solution of the quintic by radicals. We can then proceed with the equation of degree 6, choosing now rational functions that have less than six values, etc.

> But is it always possible to find such functions for equations of an arbitrary degree, that is of any number of roots? It is on this question that it is very difficult to say anything in general.
>
> (Lagrange 1770–1, §86, p. 355, in Lagrange's *Oeuvres*, Vol. 3)

Thus, in the end, Lagrange, like all his predecessors, had to give up, at least for the moment, but he had not lost all hope:

> It would be timely to apply them [the "true principles of the resolution of equations" analyzed earlier] to the equations of fifth and superior degrees whose resolution is unknown until now. But this application requires such a great number of researches and combinations, whose success is moreover still very doubtful, that we cannot for the present devote ourselves to this work. However, we hope that we can return to it another time. Here we are satisfied to have laid the foundations for a theory that seems new and general to us.
>
> (Lagrange 1770–1, §109, 403)

The frustrating lack of success was expressed by a military metaphor by Lagrange's contemporary Montucla who wrote about the problem of the solution of the quintic:

> the ramparts are raised all around but, enclosed in its last redoubt, the problem defends itself desperately. Who will be the fortunate genius who will lead the assault upon it or force it to capitulate.
>
> (translated in Ayoub 1980, 257)

In a few places in the paper, Lagrange admitted that the problem may be impossible. For example, he concluded his investigation of Euler's and Bezout's methods with the words:

> From this it follows that if the algebraic resolution of equations of degrees higher than the fourth is not impossible, it must depend on some functions of the roots different from the previous ones,
>
> (Lagrange 1770–1, §86, p. 357 in *Oeuvres*, Vol. 3)

However, he stressed that there were no clear indication of such an impossibility:

> The problem of the resolution of equations of degrees higher than the fourth is one of those for which one has not yet been able to arrive at the goal, although on the other hand nothing demonstrates its impossibility.
>
> (Lagrange 1770–1, *Oeuvres*, Vol 3, 305)

In the introduction to the paper, Lagrange emphasized the utility of his investigations for other researchers in the area:

> it will be useful for those who will work on the resolution of [equations of] higher degrees by supplying different views on this object and in particular by sparing them a great number of useless steps and attempts.
>
> (Lagrange 1770–1, *Oeuvres*, Vol. 3, 207)

It is interesting that while Lagrange here stressed the utility of his work for those (including himself) who would continue the work on the resolution of the quintic, he did not even hint at its possible use in a proof of impossibility of such a solution. Indeed, when Lagrange in the last quote talks about useless attempts it is quite clear that he speaks about the attempts to find an algebraic solution, not attempts to show its impossibility. Did Lagrange consider the possibility of such a proof at all? The only hint in this direction is his remark that "nothing demonstrates its impossibility" (see earlier quote from Lagrange 1770–1, *Oeuvres*, Vol 3, 305). If this is a reference to a possible proof of impossibility it is very hidden, and not a clear indication of a fruitful direction of research. Like many earlier mathematicians, he seems to have considered the impossibility problem as a meta-mathematical question that could bar further progress in algebra, rather than as a mathematical problem that could open new vistas in this area.

While Lagrange may have been the first to contemplate that the algebraic solution of the quintic might be impossible, Gauss and Ruffini were the first to publicly announce that they believed that it was indeed impossible. In his 1799 dissertation on the fundamental theorem of algebra, Gauss stated that he strongly believed that the quintic and higher degree equations were algebraically insoluble:

> After the labors of many geometers left little hope of ever arriving at the resolution of the general equation algebraically, it appears more and more likely that this resolution is impossible and contradictory... . Perhaps it will not be so difficult to prove, with all rigor, the impossibility for the fifth degree. I shall set forth my investigations of this at greater length in another place. Here it is enough to say that the general solution of equations understood in this sense [i.e., by radicals] is far from certain and this assumption [i.e., that any equation is solvable by radicals] has no validity at the present time.
>
> (Ayoub 1980, 262–3)

He repeated this conviction in his number theoretical magnum opus *Disquisitiones Arithmeticae* (Gauss 1801):

> Everyone knows that the most eminent geometers have been ineffectual in the search for a general solution of equations higher than the fourth degree, or (to define the search more accurately) for the reduction of mixed equations to pure equations. And there is little doubt that this problem does not so much defy modern methods of analysis as that it poses the impossible (cf. what we have said on this subject in Demonstratio nova, art. 9).
>
> (Gauss 1801, §359, 445)

Thus, Gauss believed that the solution of the quintic with radicals was impossible and he even believed that it might not be so difficult to prove. However, his promised investigations were never published.

12.2 Paolo Ruffini

The year Gauss pronounced his belief that the quintic was algebraically insoluble, a rather unknown Italian mathematician Paolo Ruffini (1765–1822)[3]

[3] For a biography see O'Connor and Robertson (1998) and Ayoub (1980). Ruffini's mathematical works were published in 1915 (Ruffini 1915).

published a book entitled: *General theory of equations in which it is shown that the algebraic solution of the general equation of degree greater than four is impossible.* At the beginning of the Introduction, he declared his debt to Lagrange:

> The algebraic solution of general equations of degree greater than four is always impossible. Behold a very important theorem which I believe I am able to assert (if I do not err): to present the proof of it is the main reason for publishing this volume. The immortal Lagrange, with his sublime reflections, has provided the basis of my proof.
>
> (translation in Ayoub 1980, 263)

Ruffini's long and difficult proof is analyzed with modern terminology by Ayoub (1980). A central element in the proof is an investigation of the number of values that a rational function of the five roots of a quintic can have. Lagrange had shown that the number of values must be a divisor of 5!=120 (Lagrange's version of his own index theorem), but Ruffini could prove that the number could not be 3 or 4. That means that one cannot find a resolvent equation of degree 3 or 4.

Ruffini sent his book to Lagrange for his opinion; when he received no answer he sent the book two more times but still received no reply. He rewrote the proof three times (publ. 1803, 1808, 1813) to make it more understandable and rigorous but except from a few Italians and one Frenchman, no one openly endorsed the result.

The opinion of Ruffini's work in Paris, the mathematical capital of the world of the time, can be sensed from the *Rapport historique sur les progrès des sciences mathématiques depuis 1789 et sur leur état actuel*, which was edited by the Secrétaire Perpetuel of the Academy of Sciences, Jean-Baptiste Joseph Delambre (1749–1822), and handed over to Napoléon in 1808 (Delambre 1810). It consisted of a consise "discours" by Delambre followed by a more detailed "Rapport" that for the pure parts followed a report written by Sylvestre François Lacroix (1765–1843).[4] In the discours, Delambre wrote:

> Mr. Lagrange's memoirs on the complete resolution of literal equations [i.e., equations with letters as coefficients], has, by reducing the problem to its least terms, shown how difficult it still is. M. Ruffini proposes to prove that

[4] Delambre (1810, 33).

> it is impossible, M. Lagrange would at least facilitate the solution of numeric equations.
>
> (Delambre 1810, 8)

According to Ayoub, Ruffini replied:

> I not only proposed to prove but in reality did prove ... and I had in mind presenting the proof to the Institute [de France] to have it examined and to have the Institute pronounce on its validity.
>
> (Ayoub 1980, 270)

In Delambre's longer report one can find the same lack of approval of Ruffini's proof. After a laudatory analysis of Lagrange's methods, Delambre (Laxroix) reported that Lagrange was now more than ever convinced of "the excessive difficulty of the problem" (Delambre 1810, 65). The report continues:

> Mr. Lagrange's learned investigations drew the geometers' attention to this point. Mr. Paolo Ruffini applies himself to proving directly the impossibility of the general solution of the problem for the litteral quantities; and returning again to the subject in volume IX of the Italian Society he sets about to determine the cases where the degree of the equation can be lowered by one which facilitates the solution and to give practical methods to effectuate the reduction when it is possible. However, since these methods are based on a difficult analysis they are not of a nature that allows them to enter in the textbooks for the primary education.
>
> (Delambre 1810, 65–6)

Having given a brief account of Lagrange's and in particular Budan's simpler numerical methods, Delambre concluded that these are

> perhaps the least incomplete that it is possible to obtain. This is at least the opinion of Mr. Lagrange, who more than any other person has the right to have an opinion on this difficult and thorny point.
>
> (Delambre 1810, 66)

Lagrange actually had a positive opinion of Ruffini's proof, but he found some holes in it. Being uncertain of the validity of the result, he did not want to approve it publically (Ayoub 1980, 271).

It is striking how little interest Ruffini's papers created in Paris. Even though they were known, few seem to have taken the trouble to read them sufficiently carefully to form an idea about their validity. No one endorsed them but no one criticized them. Only Lagrange's works were highly praised, and his opinions on the difficulty of the matter were valued more highly than Ruffini's proofs. This attitude reflects a French sense of superiority, but it may also reflect a feeling that impossibility results were less important than positive results. For example, it is characteristic that the only thing that Lacroix (Delambre) found worth evaluating in Ruffini's paper was his attempt to lower the degree of numerical equations. I think this is not only because this could be compared to a superior French method, but also because positive results, methods, and procedures leading to solutions of problems were still valued higher than negative results.

The only French mathematician who congratulated Ruffini with his proof was Cauchy. In 1821 he wrote to Ruffini:

> your memoir on the general resolution of equations is a work which has always seemed to me worthy of the attention of mathematicians and which, in my judgement, proves completely the impossibility of solving algebraically equations of higher than the fourth degree.
>
> (English translation, Ayoub 1980, 271)

Cauchy, who was otherwise sparing with acknowledging the work of his colleagues, had already in 1813 generalized Ruffini's theorem about the number of values a rational function of n variables can have under permutations of the variables. His main theorem says: The number of values of such a function cannot lie between 3 and p-1 (both included) where p is the largest prime less than or equal to n. Cauchy's paper was the first to introduce a theory of permutations independent of their use in the theory of equations. Cauchy explicitly referred to Ruffini's "remarkable results" (Cauchy 1813, 64) and in particular mentioned that Ruffini had shown Cauchy's main theorem in the case $n = 5$. However, he did not mention Ruffini's impossibility theorem concerning the solution of the quintic. Later (in 1820, see Ayoub 1980, 271), he publically approved Ruffini's result.

It is not so surprising that Cauchy should be one of the first mathematicians who publically accepted Ruffini's impossibility theorem. Indeed, Cauchy was among the first mathematicians who publically denounced the formal algebraic style of the eighteenth century, insisting on more rigorous methods, in particular in analysis. As an integral part of his new rigorous paradigm, he insisted on the importance of existence proofs. Before finding the expression of an integral (or the sum of a series, or the solution of a differential equation) he insisted that one must establish the existence of the object. Since impossibility and existence are two sides of the same coin, it is natural that Cauchy valued the importance of Ruffini's result.

12.3 Niels Henrik Abel

One of Cauchy's followers in his quest for rigor in analysis was the Norwegian mathematician Niels Henrik Abel (Figure 12.2).[5] He was also the next mathematician who announced a proof of the impossibility of the quintic. In 1821, while still a student he believed he had solved the general quintic. He sent his solution to Professor Degen in Copenhagen, who answered that though he could not find a mistake in the derivation he would ask Abel to try it out on some examples. Moreover, being of the opinion that analysis was a more important subject than algebra, Degen advised him to try his skills on elliptic integrals. Abel benefitted from both pieces of advice. He developed a revolutionary new approach to elliptic integrals, generalizing the theory to what later became called Abelian integrals, and he realized that his solution of the quintic was incorrect. Instead, he found a proof of its impossibility. He wrote up the proof in a condensed French version and published it privately, hoping it would serve as a card of entry to the international mathematical community.

Figure 12.2 Niels Henrik Abel (1802–29)
From Acta Mathematica Vol. 1

[5] The best biography of Abel is Stubhaug (2013). Skau (1990) contains a rather elementary analysis of Abel's proof. Pesic (2003) places Abel's impossibility proof in the broad context of the history of mathematics and Abel's biography. Sørensen (2004) analyzes Abel's mathematical methods more broadly.

The small pamphlet had little international impact at first, but it helped Abel win a scholarship that allowed him to visit Europe's mathematical centers. While in Berlin, he made the acquaintance of August Leopold Crelle (1780–1855), who was in the process of founding a new mathematical journal: *Journal für die reine und angewandte Mathematik.* Abel helped him with this endeavor and published a more complete and more readable version of his impossibility proof in the first volume (1826) of the journal. Soon thereafter Abel traveled to Paris where he publicized his impossibility theorem by including a long summary of it in Férussac's *Bulletin* which he also helped edit. Through these widely read venues, Abel's proof seems to have been widely known and accepted.

Abel's paper was divided into four parts. In the first part, he classified explicit algebraic functions according to how many levels of nested root signs appear in the function (the order) and how many independent root signs there are of a particular order. For example, Cardano's formula is a function of the second order in the coefficients of the equation, because there are two root signs inside each other, the outer being a third root and the inner being a square root. Using this classification Abel next proved that all the root expressions appearing in a formula for the solution of an equation must be a rational functions of the roots of the equation. This was a theorem that Ruffini had used without proof. In the third part, Abel proved Cauchy's theorem on the possible number of values that a rational function of a number of variables can have under permutations of the variables. Finally, in the last part he proved the impossibility of a general formula for the solution of the quintic. Indeed, if there exists an algebraic expression for the solution he could prove that the innermost root sign must be a square root. According to Cauchy's theorem, the next root must be either a square root or a fifth root, and Abel could prove that both assumptions led to a contradiction.

Had Abel heard of Ruffini's proofs before he found his own proof? This is a controversial question. According to Skau (1990), he did not. Yet, as argued by Ayoub (1980), it seems likely that he had at least heard of Ruffini before finalizing his pamphlet that began:

> Geometers have occupied themselves a great deal with the general solution of algebraic equations and several among them have sought to prove the impossibility. But, if I am not mistaken, they have not succeeded up to the present.
>
> (quoted from Ayoub 1980, 274).

Who could the "several" mathematicians be who had tried to prove the impossibility? Abel gave the answer himself in a manuscript written in 1828:

> The first, and, if I am not mistaken, the only one who have before me tried to prove the impossibility of the algebraic resolution of general equations, is the geometer Ruffini. However, his memoire is so complicated that it is very difficult to evaluate the correctness of his reasoning. It seems to me that his reasoning is not always satisfactory.
>
> (Abel 1828, 218)

Abel had read Cauchy's paper on permutations in which Cauchy referred to Ruffini on the first page. To be sure, Cauchy did not reveal that Ruffini had found his results on permutations in order to prove the impossibility of the quintic. Nevertheless, Abel seems to have guessed the connection.

Be that as it may, unlike Ruffini's proof, Abel's proof soon became widely known and the impossibility result was generally accepted after 1826. However, it turned out that Abel's reasoning also contained weaknesses. Already two years after its publication in Crelle's journal, Abel himself commented on the proof in the following way[6]: "I think that the proof that I have given ... leaves nothing to be desired as far as rigor is concerned. But it is not as simple as it could be" (Abel 1828, 218).

This critical appraisal of his own proof can be found in a manuscript "On the algebraic solution of equations" that Abel wrote in 1828 but did not finish during his short lifetime. It was included by Holmboe in the first edition of Abel's works published in 1839.

That same year the Irish mathematician William Rowan Hamilton (1805–65) published his discovery of a mistake in Abel's classification of algebraic functions. He concluded that "it renders it difficult to judge of the validity of his subsequent reasoning" (Hamilton 1839, 248).[7] In 1845 Wantzel referred to a wider dissatisfaction with the complexity of Abel's proof when he presented a new proof of his own, combining the best ideas from Abel and Ruffini. His paper began

[6] In a marginal note, published by Sylow and Lie in the second edition of Abel's *Oeuvres*, Abel was even more critical of his earlier proof: "Although the argument that he [the author of the present paper, i.e., Abel himself] has used seems to be rigorous, one must still admit that the method he has used leaves a lot to be desired. Thus I resume the question" (Abel 1881, vol. 2, 329).

[7] Königsberger (1869) showed how Abel's mistake could be rectified and pointed out that Abel's classification of algebraic functions was unnecessary for his proof.

> Abel has undertaken to demonstrate that an arbitrary algebraic equation of degree higher than four is not solvable by radicals. Though his demonstration is basically exact it is presented in a too complicated form which is so vague that it has not been generally accepted.
>
> (Wantzel 1845, 57)

Given these critical remarks, it is natural to ask oneself why Abel's proof was soon considered the (almost) final solution of the problem of the algebraic solution of the quintic while Ruffini's proof had been rejected or ignored. The different receptions cannot be explained by purely mathematical reasons. Both arguments had weaknesses. To be sure, Abel filled a hole in Ruffini's argument, but no one except perhaps Lagrange had bothered to read Ruffini's proof sufficiently carefully to spot this hole. Abel's greater impact was probably a result of several factors: First, Lagrange's vague unpublished comments and public lack of enthusiasm probably helped to sideline Ruffini's book and papers. Moreover, it may have helped Abel that he published his result in widely published journals and subsequently acquired a fame as a young mathematical genius. Finally, Abel's greater impact can also in part be explained as a result of changing tastes in the mathematical community. In 1800, Paris was the uncontested capital of mathematics and there, the polytechnic tradition marginalized pure mathematics. Abel published his paper in a German journal at a time when German mathematics was on the rise with its emphasis on pure mathematics in a neo-humanistic tradition. Moreover, in the period between Ruffini's and Abel's papers, mathematics had begun to move away from a formal, quantitative, and algorithmic style toward a more conceptual and qualitative style. Where an impossibility result had been considered as a lack of a result in the earlier formal Eulerian tradition, it began to acquire the status of an important mathematical result in the 1820s.

In his posthumously published paper of 1828, Abel himself pointed out that his impossibility result could be considered as a *solution* to the problem of the algebraic solution of the quintic. Having mentioned Lagrange's general methods for equation solving and their lack of success in the case of the general quintic, Abel continued:

> That led to the assumption that the solution of general equations was impossible. But this is what one could not decide since the adopted method could only lead to reliable conclusions when the equations were solvable. In fact, one proposed to solve the equations without knowing if it was possible. In

> this case, one might arrive at the solution although that was not at all certain. But if, as ill luck would have it, the solution was impossible, one could have sought for it for an eternity without finding it. Thus, in order to arrive with certainty at a result in this area, one has to take another route. One has to give the problem a form such that it is always possible to solve it, which one can do with any problem. Instead of asking for a relation in the case where one does not know whether it exists or not, one ought to ask if such a relation is in fact possible. For example, in the integral calculus, instead of searching to integrate differential formulas by groping and divination, one should rather try to decide if it is possible to integrate them in this or that way.
>
> (Abel 1828, in Abel 1881, vol. 2, 218)

This is a remarkable statement. It seems to be the first time that a mathematician so strongly emphasized the importance of impossibility results. If a problem is correctly formulated, a proof of impossibility can count as a solution of the problem. From this point of view, impossibility results become just as important as positive solutions. Moreover, Abel claimed, that if we consider impossibility proofs as solutions, then it is possible to solve any problem. This may be the earliest formulation of what Hilbert in a more formal axiomatic setting called the "axiom of the solvability of any mathematical problem" (see Chapter 16).

Abel's (and Ruffini's) impossibility theorems concerned the general quintic, i.e., the question whether there is a general formula for the solution in terms of the coefficients of the equation whatever they may be. The long search for a solution of the quintic between Cardano and Abel had revealed that many special types of equations were solvable by radicals. Therefore, having proved the impossibility of finding a general solution, Abel set himself the goal of determining all those polynomial equations that can be solved by radicals. In other words, he wanted to find a condition that the coefficients of the equation must satisfy in order for it to be algebraically solvable. He found some sufficient conditions that were more general than those of his predecessors, but it was Evariste Galois (Figure 12.3) who, in a sense, finished the job. He defined what we now call the Galois group of an equation and formulated a necessary and sufficient condition that this group must satisfy in order for the equation to be solvable by radicals. His ingenious theory was the beginning of group theory and led to the development of much of modern algebra. Galois's condition can be considered the diorism, corresponding to the question of solvability of equations.

Figure 12.3 Evariste Galois (1811–32)
Reproduced by Jim Høyer DMS

In Chapters 13 and 14, we shall see how the algebraic techniques developed in the search of a solution of polynomial equations soon became instrumental in the treatment of other questions of impossibility.

13
Constructions with Ruler and Compass
The Final Impossibility Proofs

Are the classical problems of the quadrature of a circle, the duplication of a cube, and the trisection of an angle constructible by ruler and compass? These questions, which had been raised in Greek antiquity, found their final and negative answers in the nineteenth century. We have seen that already the ancient Greeks were convinced that duplication of a cube and trisection of an angle were impossible with ruler and compass. Moreover, we have seen that in the seventeenth and eighteenth centuries Descartes and Montucla believed that they had proved the impossibility. On the other hand, the quadrature of a circle had been left wide open.

The connection of the problems to algebra had been partly revealed by new analytic techniques introduced in the seventeenth century. Descartes, in particular, had formulated the rule that problems can be constructed by ruler and compass (plane constructions) if and only if they lead to a quadratic equation. Still, his own dogma of how one ought to solve geometric problems by intersecting algebraic curves hindered a clear understanding of the exact relation between the geometry of plane constructions and the algebra of quadratic equations. The connection was formulated clearly by Gauss in 1801. Combining Gauss's insight with Lagrange's and Abel's new algebraic ideas, Wantzel (1837) could finally prove that the duplication of a cube and the trisection of an angle were impossible with ruler and compass. The quadrature of a circle turned out to be more difficult, but in 1882 Lindemann would prove that it was impossible too.

13.1 Gauss on regular polygons

While still a teenager, Carl Friedrich Gauss (Figure 13.1) made his first mathematical breakthrough when he showed how to construct the regular 17-sided polygon with ruler and compass. This was the first time a mathematician had constructed a new regular polygon with a prime number of sides

A History of Mathematical Impossibility. Jesper Lützen, Oxford University Press.
 DOI: 10.1093/oso/9780192867391.003.0013

since Euclid had shown how to construct the equilateral triangle and the regular pentagon. Gauss published his discovery in 1801 in the last chapter of his magnum opus *Disquisitiones Arithmeticae*, together with a complete characterization of all constructible regular polygons. In the earlier chapters of the book, he introduced the notation and the most important theorems in modular arithmetic and systematized the earlier works of Fermat, Euler, and Legendre on number theory. The number-theoretical results were then put to use in the study of regular polygons.

Figure 13.1 Carl Friedrich Gauss (1777–1855) and Wilhelm Eduard Weber (1804–91) constructed the first electromagnetic telegraph in 1833
Source: Archives of the Mathematisches Forschungsinstitut Oberwolfach

Gauss' point of departure was the observation that the equation

$$x^n - 1 = 0 \tag{13.1}$$

considered as an equation in the complex domain, has n roots that in the complex plane form a regular n-gon inscribed in a unit circle.[1] The polynomial $x^n - 1$ is divisible by $x - 1$ and the quotient

$$x^{n-1} + x^{n-2} + \cdots + x + 1 \tag{13.2}$$

has the same roots as $x^n - 1$ except the root 1. Assuming now that n is an odd prime, Gauss proved that this polynomial, called the cyclotomic polynomial,

[1] In fact Gauss did not explicitly introduce the complex plane, but observed that the cosine and sine of the angles of the vertices in the said n-gon will satisfy the polynomial equations that one gets by taking the real and imaginary part of the equation $x^n - 1 = 0$.

is irreducible over the rational numbers. This is one of the first appearances of the idea of irreducibility of polynomials, a concept that we saw Descartes and his followers had missed in their work. After Gauss's introduction of the idea, it became a central concept in Abel's and Galois's study of the solvability of equations.

Gauss further pointed out that the roots of the cyclotomic polynomial are powers of one of the roots. Through a study of the structure of the collection of roots and suitable combinations of them called periods, Gauss could prove the following theorem:

If $n - 1 = p_1 p_2 \cdots p_k$ is a factorization of $n - 1$ into prime factors, the solution of the cyclotomic equation (13.2) can be solved by solving a sequence of k equations of degrees $p_1, p_2, \ldots, p_k$, respectively. Here the coefficients in the first equation are rational numbers and the coefficients in the following equations are rational expressions in the roots of the previous equations. The sought-for root of the cyclotomic equation is a solution of the last of the k equations. The strategy of reducing the solution of an equation to the solution of such auxiliary equations (resolvent equations) is similar to that of Lagrange, but Gauss's methods were rather different, making little use of the theory of permutations.

At this point in his deduction, Gauss expressed his conviction that it would be impossible to find a general solution of polynomial equations by radicals (quoted in Chapter 12) and continued:

> Nevertheless, it is certain that there exist a huge number of mixed equations of any degree which allows such a reduction to pure equations, and we hope that it will not be unwanted by the geometers when we show that our auxiliary equations always belong to those.
>
> (Gauss 1801, §359)

Thus, Gauss showed that the cyclotomic equation was solvable by radicals for all prime values of the degree n. Abel's later investigations on solvable equations generalized this result.

But what about construction by ruler and compass? Gauss dealt with that in §365 of the *Disquisitiones Arithmeticae*:

> In the previous investigations we have reduced the division of the circle in n equal parts where n is a prime, to the solution of as many equations as the number of [prime] factors ... of n-1; and in fact the degrees of the equations are equal to the respective [prime] factor. Thus, when n-1 is a power of two, which is true of the following values of n: 3, 5, 17, 257, 65537, ..., the

> division of the circle can be reduced to nothing but quadratic equations and the trigonometric functions of the angles $\frac{P}{n}$, $\frac{2P}{n}$, ...[2] can be expressed through more or less complicated (according to the value of n) quadratic equations. In these cases the division of the circle in n equal parts or the description of a regular polygon with n sides can therefore be done by geometric constructions.
>
> (Gauss 1801, §365)

With this argument Gauss had established the remarkable theorem that a regular polygon with n sides, where n is a prime, is constructible by ruler and compass if n-1 is a power of 2 or equivalently if n is a so-called Fermat prime, that is, a prime of the form $2^l + 1$. It is easy to prove that if a number of this form is a prime, the exponent l must itself be a power of 2 so that the number is of the form $n = 2^{2^m} + 1$. The numbers mentioned by Gauss in the quote are the values of $2^{2^m}+1$ for $m = 0, 1, 2, 3, 4$, which are indeed primes. Fermat had famously conjectured that all numbers of the form $2^{2^m} + 1$ are prime, but Euler proved that for $m = 5$ the number $2^{2^m} + 1$ is not a prime. Since then one has tried to find more Fermat primes than the five primes mentioned by Gauss but without success. On the other hand, no one has been able to prove that there does not exist another Fermat prime. So, the dots in Gauss's list of Fermat primes cover an enigma.

The above quote from the *Disquisitiones Arithmeticae* also implicitly contains the first translation into algebra of the geometric concept of constructibility with ruler and compass. Indeed, Gauss implied that if

> the principal unknown of a [geometric construction] problem can be obtained by the resolution of a series of quadratic equations whose coefficients are rational functions of the givens of the problem and the roots of the previous equations.
>
> (Wantzel 1837, 566)

then the construction can be done by ruler and compass. Gauss made no particular point out of this algebraic characterization of constructible problems, and seems to have considered it trivial. Indeed, it is a trivial consequence of the fact, known implicitly to the ancient Greeks and explicitly pointed out by Descartes, that a quadratic equation can be solved by ruler and compass. Still, we have seen that all of Gauss's predecessors had been unable to algebraically mirror the successive nature of ruler and compass constructions, so Gauss's implicit formulation was a great step forward.

[2] P is Gauss's name for the entire circle or 2π.

The explicit formulation quoted above of the algebraic criterion for solvability by ruler and compass was formulated by Pierre Wantzel (1814–48). However, Gauss and Wantzel formulated converse implications: Gauss wrote (implicitly) that *if* a geometric problem can be solved by the solution of a succession of quadratic equations, *then* it is constructible by ruler and compass. He needed this implication for his proof that polygons with a prime number of sides are constructible if the number of sides is a Fermat prime. Wantzel, on the other hand, wanted to prove that no other regular polygons with a prime number of sides can be constructed by ruler and compass. Thus, he formulated the converse. The converse is true as well, but it is less obvious. Wantzel gave an eight-line-long argument that can hardly be considered complete. Instead, it seems to have been the Danish mathematician Julius Petersen (1839–1910) who first gave a satisfactory argument for the equivalence of the algebraic formulation and the geometric property of constructibility by ruler and compass. He first published the proof in Danish in his doctoral dissertation (1871), and then in his algebra book (1877) that was translated into French and German. The argument was popularized by Felix Klein (1849–1925) in a book of 1895. Instead of successive solutions of quadratic equations, Petersen phrased himself in terms of successive adjunction of square roots in the style of Abel. But the result is the same.

Petersen pointed out that the coordinates of a point of intersection of two circles given by their radii and the coordinates of their centers can be expressed rationally in terms of the given coordinates and radii using at most one square root (of the discriminant of the equation). The same holds for the intersection of a circle and a line, while the intersection of two lines can be expressed rationally in the coefficients. Thus, consecutive use of ruler and compass can only lead to numbers expressed in terms of nested square roots.

In the context of the present book, it is most relevant to emphasize that Gauss also claimed that it is impossible to construct regular polygons with a prime number of sides when this number is not a Fermat prime.

> However, when $n-1$ contains other prime factors than 2, we will always arrive at higher equations namely to one or more cubic equations when 3 occurs once or several times among the prime factors of $n-1$, to equations of the fifth degree when $n-1$ is divisible by 5 etc. *and we can prove with all rigor that* **these higher equations cannot be avoided or reduced to equations of lower degree.** Although the limits of this book does not permit to provide this proof here, we still thought that we must call attention to it so that no one would imagine that he can reduce the divisions other than the ones given by

> our theory, e.g. in 7,11,13,19, ... parts to geometric constructions and waste his time.
>
> (Gauss 1801, §365)

Thus, Gauss claimed that he could prove the impossibility of constructing the other polygons with a prime number of sides, but did not include his proof, because he did not want his book to grow too big. Gauss later became notorious for such claims of priority. Several of his claims were challenged by Legendre but as far as I know, no one disputed that Gauss knew a proof of the mentioned impossibility. Indeed, as proved later by Wantzel, the impossibility proof is simpler than Gauss' constructive proof of the converse and only needs techniques developed by Lagrange and Gauss himself.

Today the impossibility part of Gauss' result is just as famous as the positive converse. The fact that Gauss chose to leave out the proof of the impossibility claim indicates that for him impossibility results were still less important than constructive results.

On the basis of his result concerning constructibility of regular polygons with a prime number of sides, Gauss could rather easily characterize all constructible regular polygons: A regular polygon with n sides is constructible with ruler and compass, if and only if n is a product of a power of 2 and a number of *different* Fermat primes.

If Gauss had proved the impossibility part of this theorem, he would also have shown the impossibility of the trisection of an arbitrary angle. Indeed, as proved by Euclid, one can construct a regular triangle having all angles equal to 60°, but it is impossible to construct a regular polygon with nine sides, since $9 = 3 \cdot 3$ and thus a product of two *equal* Fermat primes. Therefore, one cannot trisect an angle of 60°. As it were, the impossibility claims made by Gauss were only proved 35 years later.

13.2 Wantzel

In a paper of 1837 in the second volume of Liouville's *Journal de Mathématiques pures et appliquées*, Pierre Laurent Wantzel finally provided a proof of the impossibility part of Gauss's theorem concerning constructible regular polygons as well as a proof of the impossibility of constructing the duplication of a cube and the trisection of an angle. More generally, the paper was concerned with "Researches on the ways to recognize if a geometric problem can be solved by ruler and compass." The main theorem of the paper stated that if a problem is constructible its main unknown is a solution of an equation of degree 2^n which is irreducible over the rational numbers to which has been adjoined possible known quantities of the given problem.

Taking his point of departure in the system of quadratic equations that characterize a constructible geometric problem (quoted earlier), Wantzel unpacked the quadratic equations starting with the last one and arrived at a polynomial of degree 2^n with coefficients that are rational in the givens of the problem (for details see Lützen 2009). He then claimed that if the number of quadratic equations has been reduced as much as possible, the resulting equation is irreducible. He used methods reminiscent of those employed by Abel in his investigation of the quintic. The proof of the irreducibility of the equation is rather convoluted, and as pointed out by Hartshorne (2000) (but not earlier!), there is a gap in it that was only circumvented by Petersen (1871, 1877).

Since the duplication of a cube depends on the solution of the irreducible equation $x^3 = 2$ whose degree is not a power of 2, this classical problem is not solvable by ruler and compass. In the same way Wantzel claimed that the cubic equation that one must solve in order to trisect an angle (Section 7.3) is irreducible, at least for certain values of the given angle, so that also the trisection of an angle is impossible in general. He was justly proud of his achievements:

> It seems to us that it has not been demonstrated rigorously until now that these problems, so famous among the ancients, are not capable of a solution by the geometric constructions they valued particularly.
>
> (Wantzel 1837, 369)

Wantzel continued with a brief and elegant proof of the converse of Gauss's theorem about constructible regular polygons that he formulated as an impossibility theorem:

> The division of the circle into N parts cannot be done by ruler and compass unless the prime factors of N different from 2 are of the form $2^n + 1$ and they enter only in the first power in this number.
>
> (Wantzel 1837, 369)

One would have imagined that Wantzel would have become instantly famous for having finally settled two of the most famous ancient problems of geometry. However, that did not happen. His paper was hardly noticed when it was published and in 1894, in a volume celebrating the centenary of the École polytechnique, where Wantzel had worked, Lapparent remarked: "Wantzel, in the eyes of the world, is forgotten" (Lapparent 1895, 133). Petersen, who seems to have been the next mathematician who proved the impossibility of the two classical problems, referred to Wantzel in his doctoral dissertatin in Danish (Petersen 1871, 44). However, when he published his proof in his algebra book (1877), the reference was left out and thus many authors of the late nineteenth century believed that he had been the first to prove the impossibility. Other authors attributed the result to Klein, who had made Petersen's proof more accessible in 1895.

Wantzel's accomplishments were rediscovered by Cajori in 1918. In a paper on the life and work of this forgotten French mathematician, he stated:

> Quite forgotten are the proofs given by Wantzel of three theorems of note, viz. the impossibility of trisecting angles, of doubling cubes, and of avoiding the "irreducible case" in the algebraic solution of irreducible cubics. For these theorems Wantzel appears to have been the first to advance rigorous proofs.
>
> (Cajori, 1918, 345)

Beginning with Tropfke's history of elementary mathematics (1937, 125) it began to be commonplace to attribute the theorem to Wantzel. Why did it take a century before Wantzel's proof was finally recognized? Let me start with three non-reasons. As I pointed out earlier, there was a subtle hole in Wantzel's proof. However, since this hole was not noticed until Hartshorne pointed it out in 2000, it clearly had no influence on the reception of the result. Secondly, the journal in which Wantzel published his proof was, next to *Crelle's Journal*, the most highly ranked journal of the time, so the missing recognition was not due to a small impact of the venue of publication. Finally, the result was not rejected because it was controversial or marginal. Indeed the classical problems were among the most celebrated ancient problems, and their impossibility had been expected for about 2000 years.

The last of the non-reasons may be closely linked with one of the reasons for the missing recognition: As we saw in Section 7.5, the French Academy had in 1775 publicly stated that they would no longer examine solutions of these classical problems. They had not banned proofs of impossibility, but their rhetoric had probably given the problems a bad reputation, and more importantly, in the motivation of the decision, the academy had claimed that the impossibility of the duplication of a cube and the trisection of an angle were proven facts. It is therefore likely, that many mathematicians at the beginning of the nineteenth century still believed that Montucla's argument more or less proved the impossibility. So for them Wantzel's result would be old news. Yet, the lack of recognition also seems to stem from a lack of interest in such impossibility claims. In particular, in France, the eighteenth-century constructive paradigm was still alive, and we have seen that even Gauss, the Princeps Mathematicorum as he was called, had not valued the impossibility proof so highly that he would waste pages of his book on it.

To be sure, a number of young mathematicians had begun to show an interest in impossibility questions. We have already mentioned Abel and Galois, but also Liouville, the publisher of Wantzel's paper (see Chapter 14), belongs to this group. Wantzel, himself was keenly interested in problems of impossibility. In addition to the impossibility of two of the classical problems and the construction of regular polygons, he also published a revised proof on the impossibility of solving the general quintic by radicals, as well as a related result mentioned by Cajori in the earlier quote. The result concerns the cubic equation: Cardano's formula gives a solution by radicals, but in the so-called irreducible case, the radicals give rise to complex numbers even though the roots of the equation are real. The question is whether one can find another expression of the roots in terms of real radicals. Wantzel proved that this is impossible.

In the only original biography ever written of Wantzel, the author Barré de Saint-Venant (1848) asks why Wantzel accomplished so little. He informs his readers that some have ascribed it to "the metaphysical form of his mind." Is it possible that these unnamed persons thought that Wantzel's attraction to impossibility results was the result of a metaphysical mind? If so, it would indicate that impossibility results were still considered as meta-mathematical statements rather than proper mathematical results.

13.3 The quadrature of a circle

The last of the classical problems, the quadrature of a circle, held out another half-century. From the ancient Greeks through the eighteenth century it was in general considered unknown whether the problem could be solved by ruler and compass. However, from the time of Archimedes, it was known that the constructibility of the quadrature of a circle was equivalent to determining the nature of the ratio of the circumference of a circle to its diameter, a ratio that was gradually considered as a number and eventually named π in the eighteenth century. In Section 9.8 we saw that in 1768 Johann Heinrich Lambert proved that π was irrational. Lambert's result clarified the nature of π, but it did not decide the question of the quadrature of a circle. Indeed, all irrational numbers that can be expressed by square roots are constructible. Still, with Gauss's and Wantzel's results it became clear that all constructible numbers are roots of polynomial equations with rational coefficients. Lambert had claimed as much, without proof. Moreover he had claimed (without proof) that π is a transcendental number, i.e., not a root of *any* polynomial equation with rational coefficients. From these unproven claims he had concluded that the quadrature of a circle is impossible with ruler and compass.

However, as late as 1840 no one had been able to show that there exist transcendental numbers at all. The first to do so was Joseph Liouville (1809–82). In an 1844 paper he approached the question using continued fractions but in 1851 he formulated his theorem in the following more general terms: If a sequence of rational numbers converges to an irrational limit faster than any power of the inverse of its denominators, then this limit is a transcendental number. As an example he mentions the sum of the series:

$$\frac{1}{l} + \frac{1}{l^{2!}} + \frac{1}{l^{3!}} + \cdots + \frac{1}{l^{n!}} + \cdots,$$

where l is a natural number larger than 1. His notebooks reveal that his discovery was a result of a failed attempt to prove that the basis of the natural logarithm, e, is transcendental. At some point he also believed that his theorem had the consequence that π is transcendental; however, he soon discovered that his argument was flawed.[3]

Liouville's student Charles Hermite (1822–1901) had better luck. In 1873 he published in the *Comptes Rendus* of the Paris Academy a proof that e is transcendental. Some of his German colleagues urged him to try out his methods on π as well, but in a letter to Borchardt who was the editor of *Crelle's Journal* at the time he replied:

> I will not venture to search for a demonstration for the transcendence of π. If others attempt that enterprise, no one will be happier than me should they succeed, but believe me, my friend, it will cost them some real effort.
>
> (Hermite 1873b, 342)[4]

Nevertheless, nine years later, the otherwise rather unknown German mathematician Ferdinand Lindemann (Figure 13.2) was able to prove that π is

[3] For more information on Liouville's work on transcendental numbers see Lützen (1990, Chapter 12).

[4] The translation of the quote is by David Rowe, whose paper (Rowe 2015) tells the interesting story connecting Lindemann's discovery with Hilbert's seventh Paris problem.

transcendental. In that way he had finally establish that the quadrature of a circle is impossible with ruler and compass. He explicitly acknowledged his debt to Hermite:

> The most essential basis of the investigation are the relations between certain definite integrals that Hermite has used to establish the transcendental character of the number e.
>
> (Lindemann 1882, 213)

With Hermite's methods, Lindemann could show that if z is an algebraic real or complex number different from 0, then e^z is irrational. Since $e^{i\pi} = -1$, Lindemann concluded that $i\pi$ is transcendental from which it follows that π is transcendental.

Figure 13.2 Ferdinand Lindemann (1852–1939)
Source: Archives of Peter Roquette and Archives of the Mathematisches Forschungsinstitut Oberwolfach

Lindemann sent his paper to Felix Klein for publication in the *Mathematische Annalen*. Klein asked Georg Cantor (1845–1918) to review the paper. Cantor was soon persuaded that the proof was essentially correct although some details needed to be worked out. He also pointed out that Lindemann's result implied that the graph of the exponential function went through the real plane without hitting any points with two algebraic coordinates (except (0,1)). This highlighted Cantor's result from 1870 to the effect that there are immensely many more transcendental numbers than algebraic numbers. To be more precise, there are only countably many algebraic numbers, whereas the transcendental numbers cannot be counted (see Section 16.4).

While waiting for the publication of the paper, the leading Berlin mathematician Weierstrass, who Cantor had informed about the paper, offered to announce Lindemann's result to the Berlin Academy so as to secure Lindemann's priority of the remarkable result. That happened in June 1882 and later the same year, *Mathematische Annalen* published Lindemann's 13-page paper. It ended with a generalization of the main theorem of the paper, namely:

If z_i $(i = 1, 2, \ldots, r)$ are different algebraic numbers and N_i are algebraic numbers that are not all equal to 0, then the following equation cannot hold:

$$N_1 e^{z_1} + N_2 e^{z_2} + N_3 e^{z_3} + \cdots + N_r e^{z_r} = 0.$$

He only offered what he himself called a rudimentary proof of this theorem and concluded: "I reserve the right to give a more detailed explanation of the evidence that is only hinted at here for a later publication" (Lindemann 1882, 225).

However, Lindemann never published the promised detailed proof and indeed he never published another important paper in his life.[5] Instead other mathematicians took over. Weierstrass soon found a rigorous proof of the last general theorem in Lindemann's paper, and for that reason it is known as the Lindemann–Weierstrass theorem. Moreover he simplified Lindemann's entire argument and made it more rigorous. He published his simplified proof in 1885, after having given Lindemann a fair chance to publish his promised detailed proof.

Weierstrass' proof was in turn simplified further by Dedekind, H. A.Schwarz, and Hurwitz through whom Hilbert became interested in the question of transcendental numbers. Hilbert, in turn, published a famous proof of only four pages that was also inspired by a paper by Stieltjes (1890) (see Rowe 2015, 224–30). Hilbert's proof was popularized through Felix Klein's small book *Famous Problems of Elementary Geometry* (Klein 1895). As mentioned in Section 13.1, this book also contained a brief account of Petersen's proof (1877) of the impossibility of doubling a cube and trisecting an angle by ruler and compass. Klein's book did a great job of spreading the news

[5] Instead he tried his hand at Fermat's last theorem. According to Rowe (2015), "his numerous claims to have solved it during the remaining decades of his long life eventually made him appear like a laughing-stock to those who had some inkling of serious research in number theory" (Rowe 2015, 225).

of the impossibility of the three classical problems, closing a more than two-millennia-long quest for their solution.

So, in contrast to Wantzel, whose results were not noticed for almost a century, Lindemann became instantly famous for his proof of the impossibility of the last of the classical construction problems. This indicates that impossibility theorems were finally beginning to be accepted as important mathematical results.

14
Impossible Integrals

14.1 Early considerations

In the chapters on the classical problems, we saw that curves were classified into groups of increasing complexity. According to Pappus, a straight line and a circle are the simplest, followed by conic sections. All other curves make up the last and most complicated class of "lines." While solving problems, simple curves are preferred over more complicated ones. Descartes refined the last class, dividing them into algebraic curves, which are ordered according to the degree of the defining polynomial, and the remaining "mechanical" curves, which Descartes banned from geometry altogether. Next, Leibniz accepted mechanical curves that he called transcendental, emphasizing proudly that his new differential calculus was capable of dealing with them as well.

With Euler, the classification of curves was transformed into a classification of functions, which he considered to be analytic expressions in a variable x and a number of constants. The expressions containing only the rational operations $+, -, \cdot, :$ were called rational functions. If root signs or radicals were needed, the function was called (explicit) algebraic. Solutions of a polynomial equation were sometimes considered (implicit) algebraic functions of the coefficients, even if the solution was not known in terms of radicals. Euler also introduced a number of standard transcendental functions. The logarithms and exponential functions, as well as the trigonometric functions and their inverses, were considered the simplest ones. If a function can be written as a finite expression using only algebraic operations, exponentials, logarithms, and trigonometric functions, it is often said to be expressed in finite or elementary terms. If we allow complex numbers, we can leave trigonometric functions and their inverses off the list because they can be expressed in terms of exponentials and logarithms, according to Euler's formulas.

In addition to the elementary transcendental functions Euler and his contemporaries encountered other more complicated transcendental functions. They typically arose as integrals. From the start of the calculus, it was noted that differentiation is easier than integration. Differentiation of simple functions can always be expressed in terms of equally simple functions. Integration,

A History of Mathematical Impossibility. Jesper Lützen, Oxford University Press.
 DOI: 10.1093/oso/9780192867391.003.0014

on the other hand, often leads to more complicated functions. For example, integration of the algebraic function $\frac{1}{x}$ leads to a transcendental function, the logarithm. Similarly, Euler encountered many integrals of algebraic or elementary functions that he could not express in finite terms. A famous class of such integrals is the so-called elliptic integrals that are of the form

$$\int \frac{P(x)}{\sqrt{R(x)}} dx,$$

where $P(X)$ is a rational function and $R(x)$ is a polynomial of degree 3 or 4.

In general, Euler and his successors would prefer to express an integral in finite terms if possible. They devised many tricks to determine the value of integrals. The French mathematicians Alexis Fontaine (1704–71) and Marie-Jean Marquis de Condorcet even claimed that they had found universal methods for solving the more general problem: solve any "solvable" differential equation in algebraic or elementary terms. Fontaine tabulated long lists of possible algebraic solutions to various types of differential equations, but as Condorcet pointed out, he did not limit the possible solutions to a finite number of types, such that his determination of algebraic solutions became unmanageable. Liouville expressed the problem more pointedly:

> The general method indicated by Fontaine is in reality nothing but a laborious groping whose least fault is its disheartening length. These gropings do not constitute a regular procedure for the reason that no circumstance tells us after how many attempts one will arrive at the desired result i.e. the integral that one wants to calculate. Thus, if by chance this integral is impossible in the algebraic form that one a priori assigns to it, one will never be warned, however far one pushes these operations.[1]
>
> (Liouville 1834, 37–8)

Condorcet went further than Fontaine, trying to limit the possible algebraic or finite expressions of the solutions to differential equations to a manageable number of possibilities. However, his most wide-ranging ideas were not published,[2] he did not provide rigorous proofs, and in the end, he was unable

[1] It is interesting to compare Liouville's formulation with Abel's very similar critique quoted in Chapter 13 of the earlier attempts to solve the quintic by radicals. Abel's critique was written earlier than Liouville's, but it was published later.

[2] Condorcet's unpublished works on integration in finite form has been investigated by Gilain (1988).

to show that any particular integral could *not* be expressed in algebraic or finite terms.

In his book on probability theory (1812), Pierre-Simon Laplace (1749–1827) emphasized that "the exponential, algebraic and logarithmic functions cannot be reduced to one another" (Laplace 1812, 4) and moreover that different radical expressions cannot be reduced to one another:

> These principles, which are based on the very nature of functions, can be of great use in analytic research because they indicate which form the functions we are searching for must necessarily have, and show that it does not exist in a great number of cases .... In this way, I have realized that one cannot obtain the integral $\int \frac{dx}{\sqrt{1+\alpha x^2+\beta x^4}}$ as a finite explicit or implicit function.
>
> (Laplace 1812, 5)

Laplace also claimed that he had a general method for finding solutions of differential equations in finite terms, when such solutions exist, and that he could prove the impossibility if such solutions do not exist. However, he did not reveal his methods.

14.2 Abel's mostly unpublished results

In Chapter 12 we saw that Abel mentioned the problem of integration in finite terms when he discussed how one ought to pose problems:

> For example, in the integral calculus, instead of searching to integrate differential formulas by groping and divination one should rather try to decide if it is possible to integrate them in this or that way.
>
> (Abel 1828 in Abel 1881, vol 2, 218)

While still a student, Abel wrote a paper on integration in finite terms. It was never published and seems to be lost but some of its content made it into his later papers on elliptic and abelian functions (see Ore 1974, 63 and Lützen 1990, 359–69). For example, in a letter to Legendre published 1830, Abel claimed that "with the aid of a particular method" he had proved the following theorem:

> If an integral, $\int y\partial x$ where y is related to x by an arbitrary algebraic equation, can be expressed in some way explicitly or implicitly by way of algebraic and logarithmic functions, one can always suppose that:
>
> $$\int y\partial x = u + A_1 \log v_1 + A_2 \log v_2 + \cdots + A_m \log v_m,$$

> where A_1, A_2,... are constants and u, v_1, v_2,..., v_m are rational functions of x and y.
>
> (Abel 1830, *Oeuvres* II, 275–6)

Abel published a proof of a generalization of this theorem in his *Précis d'une théorie des fonctions elliptiques* (Abel 1829, *Oeuvres* I, 549–50). In conformity with his work on the solution of equations by radicals, his proof of the theorem on integration of algebraic functions was based on Lagrange's ideas on permutations and his own ideas on irreducible polynomials.

He used the theorem stated above to investigate when an integral of the form $\int \frac{\rho dx}{\sqrt{R}}$, where ρ and R are polynomials, can be written in finite form. His method relies on a continued fraction expansion of $\sqrt{R}$ but did not lend itself to proofs of impossibility. Still, in the earlier-mentioned unfinished paper of 1828 Abel managed to prove that the elliptic integral $\int \frac{x^2 dx}{\sqrt{R}}$, where R is a polynomial of degree 4, cannot be expressed in algebraic form or in logarithmic form.

According to Jacobi, who was Abel's rival in the field of elliptic functions, Abel's contributions to the problem of integration in finite terms was peculiar to him:

> The wide scope of problems he proposed for himself, to find a necessary and sufficient condition for an algebraic equation to be solvable by radicals, for an arbitrary integral to be expressible in finite form, his wonderful discovery of a general property shared by all functions which are integrals of an algebraic function, etc., etc.—all these are questions of a form which is peculiar to him; no one before him had dared to propose them.
>
> (Ore 1974, 233–4).

Before the publication in 1839 of Abel's last-mentioned impossibility proof, Liouville had taken the question of integration in finite terms much further.

14.3 Joseph Liouville on integration in algebraic terms

The first mathematician who published widely on integration in finite terms was the young Frenchman Joseph Liouville (Figure 14.1). The notes in his still preserved notebooks bear witness to the gradual development of the theory (see Lützen 1990, 369–422). In the fall of 1832 he began to investigate when the integral of an algebraic function is again algebraic. In his first note, he proved that the elliptic integral $\int \frac{dx}{\sqrt{1+x^4}}$ is not an algebraic function. His argument used a mixture of algebraic and analytic techniques. The algebraic methods were inspired by Lagrange and Abel; the analytic methods had to do with the value of the integral at particular values, its parity, and in particular its growth at infinity. From this particular integral, Liouville consciously embarked on a sequence of generalizations to more and more general types of integrands. In the process, the analytic arguments gradually disappeared and the proofs became almost completely algebraic. In the beginning, Liouville took algebraic functions to mean explicit algebraic functions (i.e., explicit expressions in radicals) that he classified similarly to Abel's classification in his paper on the impossibility of the quintic. After a few months, however, he generalized the concept to cover also implicit algebraic functions defined by polynomial equations that may or may not be solvable by radicals. This step may have been inspired by Abel's papers on integration. Through his gradual generalizations, Liouville succeeded in solving the problem of integration in algebraic terms completely.

He published his solution in three papers (Liouville 1833a (in two parts) and 1833b). The main trick was to transform the problem into the problem of finding rational solutions to a linear differential equation with rational functions as coefficients. This problem could, in turn, be transformed into finding polynomial solutions to a linear differential equation with polynomial coefficients, and this problem can be solved by Newton's so-called analytic parallelogram if it has a solution. If it does not have a solution, the original integral is not algebraic. In this way Liouville had found a finite algorithm that can determine whether a given integral of an algebraic function is algebraic, and determine an algebraic expression of the integral when it exists.

Figure 14.1 Joseph Liouville (1809–82)
Bibliothèque de l'Institut de France

14.4 Liouville on integration in finite terms

Liouville's papers were received very positively by Siméon Denis Poisson (1781–1840), who reviewed them on behalf of the French Academy of Sciences. Still, Poisson suggested that Liouville should extend the investigations:

> by limiting his considerations to algebraic integrals of differential formulas, Mr. Liouville has therefore not completely solved the problem concerning the absolute possibility or impossibility of their integration in finite terms.
>
> (Poisson 1833, 211)

It seems most likely that from the outset Liouville had already planned a generalization of his investigations in this direction. Such a generalization had been strongly suggested by Laplace whose statements about integration in finite terms were known to Liouville, and by Abel's papers on integration that Liouville read at the latest in the beginning of 1833. Whoever or whatever inspired Liouville, he did take up the problem of integration in finite terms in the spring of 1833. His results were published the following years (Liouville 1834 and 1835). They contain the first complete proofs of the impossibility of expressing certain integrals in finite terms. The proofs rested on a classification of elementary functions, or finite explicit functions as he called them:

A function $y(x)$ is called algebraic if it is a root of a polynomial equation

$$y^n + A_{n-1}(x)\,y^{n-1} + \cdots A_1(x)\,y + A_0(x) = 0, \tag{14.1}$$

where $A_0, A_1, \ldots, A_{n-1}$ are rational functions of x. Logarithms or exponentials of algebraic functions are called (transcendental) monomials of the first kind. A function that is not algebraic is called a transcendental function of the first kind (or order) if it is an algebraic function of x and transcendental monomials of the first kind, i.e., if it is a solution of a polynomial equation (14.1) with coefficients A_i which are rational functions of x and transcendental monomials of the first kind. For example,

$$e^{x^2+3} - \sqrt[3]{\log(x+5)+1}$$

is a transcendental function of the first kind. Logarithms and exponentials of functions of the first kind are called monomials of the second kind, and algebraic functions of algebraic and first-order functions and monomials of the second kind are called transcendental functions of the second kind (unless they can be reduced to a lower kind). For example, $x + log(logx))$ is a transcendental function of the second kind. Continuing recursively in an obvious way, Liouville defined what is understood by a transcendental function of the *n*th kind. The collection of all these functions is what Liouville called explicit finite functions and what we today often call elementary functions.[3]

With this classification at hand, Liouville could prove a lemma that Ritt (1948, 16) has called Liouville's principle:

> If the number of monomials of the *n*th kind entering a function of the *n*th kind is reduced to a minimum, any algebraic relation in these monomials and functions of a lower kind must be a trivial identity, i.e., an equation that is fulfilled whatever is substituted instead of the monomials.
>
> (Liouville 1837/8, 76)

The proof of the lemma is rather obvious but the consequences are far-reaching. Indeed Liouville could apply it to prove the fundamental theorem.

Liouville's theorem: If $y(x)$ is an algebraic function and if $\varphi(x) = \int y dx$ is a finite explicit function, then it is of the form

$$\varphi(x) = \int y dx = t + A \log(u) + B \log(v) + \cdots C \log(w), \qquad (14.2)$$

where $t, u, v, \ldots, w$ are algebraic functions and $A, B, \ldots, C$ are constants.

I shall indicate the beginning of the proof, to give an impression of Liouville's methods. Assuming first that the integral is of the first kind and contains an exponential $\theta = e^u$ where u is an algebraic function, he wrote the integral on the form

$$\int y dx = \varphi(x, \theta).$$

He differentiated this equation and obtained

$$y = \varphi'_x(x, \theta) + \varphi'_\theta(x, \theta)\,\theta u'.$$

Since differentiation of an elementary finite function does not introduce new transcendental monomials, this equation must, according to Liouville's

[3] Remark that in a finite explicit function the algebraic functions can be defined implicitly through algebraic equations. But the transcendental functions, i.e., the logarithms and exponentials, can only enter explicitly.

lemma, be a trivial identity in θ. Therefore, it is still valid if we substitute $\mu\theta$ instead of θ and since y is algebraic it does not change under this substitution, so we get

$$\varphi'_x(x,\theta) + \varphi'_\theta(x,\theta)\,\theta u' = \varphi'_x(x,\mu\theta) + \varphi'_{\mu\theta}(x,\mu\theta)\,\mu\theta u'.$$

Integrating this equation from b to x and assuming $\theta(b) = a$ Liouville got

$$\varphi(x,\mu\theta) = \varphi(x,\theta)\,\varphi(b,\mu a) - \varphi(b,a).$$

Again this is an algebraic equation in the transcendental monomials contained in φ, and so it must hold if an arbitrary letter ζ is substituted for θ. Differentiating the resulting equation and setting $\mu = 1$ Liouville found that

$$\varphi'_\zeta(x,\zeta) = \frac{a\varphi'_\zeta(b,a)}{\zeta}.$$

Finally, he integrated this equation with respect to ζ from a fixed value ζ_0, reintroduced the value $\zeta = \theta = e^u$, and obtained

$$\varphi(x,\theta) = a\varphi'_a(b,a)\left(u - \log(\zeta_0)\right) + \varphi(x,\zeta_0).$$

Since the right-hand side does not contain the monomial θ, Liouville concluded that the integral could not contain any exponentials.

With similar arguments, Liouville then proved that if the integral is of the first kind, the logarithmic monomials must enter in a linear fashion as in (14.2). He then proved that if the integral is of the second kind, it must be of the form (14.2) where $t, u, v, \ldots, w$ are of the first kind, and then proved by contradiction that $t, u, v, \ldots, w$ cannot contain neither exponential nor logarithmic monomials. Thus, the integral cannot be of the second kind. This argument can easily be generalized to higher kinds of functions. This completes Liouville's proof of his theorem.

As mentioned earlier, Abel had already proved that the algebraic functions $t, u, v, \ldots, w$ in (14.2) were rational functions in x and y. Having thus limited the possible finite expressions for an integral of an algebraic function to a manageable number of alternatives, Liouville felt confident that he could find a finite algorithm to decide whether an algebraic function can be integrated in finite terms and determine the integral if it exists. However, when he tried his hand on integrals of the special kind

$$\int \frac{Pdx}{\sqrt{R}},$$

where P and R are polynomials, he realized that he was unable to deal with this type of integrals in general (Lützen 1990, 385). Still he was able to deal with two cases (Liouville 1834):

a. When R has odd degree with only simple roots, the integral is expressible in finite terms if and only if it is algebraic.
b. If R has even degree with only simple roots and $\deg(R) > 2(\deg(P)) + 2$, then the integral cannot be expressed in finite terms.

The latter result includes what Legendre had called the elliptic integral of the first kind. With a slight modification of his method, Liouville also showed the impossibility of expressing the elliptic integral of the second kind. In this way these impossibilities, which had been taken for granted by Euler, were finally proved.

Liouville soon discovered that he could use his methods to deal with integrals of the form $\int e^x y dx$, where y is algebraic. For example, he could prove that the integrals

$$\int \frac{e^x}{x} dx \text{ and } \int_0^\infty \frac{\sin(\alpha x)}{1+\alpha^2} d\alpha$$

are not elementary (Liouville 1835).

Two years later Liouville turned to the solution of transcendental equations in finite terms. More precisely, let $T(x, y)$ be an explicit finite function; Liouville described a method for determining whether the solution $y(x)$ of the equation $T(x, y) = 0$ is expressible as an explicit finite function. For example, he could prove that the equation

$$log(y) = F(x, y)$$

has no explicit finite solution $y(x)$ when F is algebraic and $F'_x \not\equiv 0$ and $F'_y \not\equiv 0$. This implies that Kepler's equation

$$x = y - h\sin(y)$$

has no finite explicit solution $y(x)$.

In Chapter 8 we saw that Newton had used the transcendental nature of an indefinite circle quadrature to argue for the impossibility of solving the Kepler problem algebraically. The above theorem strengthens this claim. Even elementary transcendental functions do not help.

The impossibility of solving Kepler's equation in explicit finite terms had been conjectured already in the eighteenth century, for example by Lambert (Lambert 1767, 355), but as Liouville remarked: "it was easier to announce this theorem than to prove it" (Liouville 1837/8, 58).

If a function $y(x)$ is defined as the solution of the equation $T(x, y) = 0$, where T is a finite explicit function, we could consider it an implicit finite function. The above theorem shows that the class of implicit finite functions is larger than the class of explicit finite functions. Liouville now asked himself: Can we integrate more algebraic functions if we also allow the integral to be an *implicit* finite function. One can find his answer in one of his notebooks:

> Let $\int y dx = z$, y algebraic, z not expressible in explicit finite form. I say that z cannot be a finite implicit function either.[4]

Thus, we cannot integrate more functions when we allow implicit finite functions. Liouville gave a formally correct proof of this theorem (Lützen 1990, 395–401) but he never published it. Therefore, the theorem is now named after Joseph Fels Ritt, who first conjectured it in public (Ritt, 1948, 94), and Robert Henry Risch, who first published a proof of it (Risch 1976).

14.5 Liouville on solution of differential equations by quadrature

Already in 1833 Liouville began to extend his research to solutions of differential equations in finite terms. His purpose was evidently to find a rigorous replacement for Condorcet's speculations. However, Liouville encountered many problems and eventually gave up the grand plan. Yet in 1839–41 he published three interesting papers containing some of his partial results. In the first paper (1839) he discussed the second-order linear differential equation

$$\frac{d^2y}{dx^2} = P(x)\, y,$$

where P is a polynomial. He could prove that it had no algebraic solutions and that the general solution was never expressible in finite terms. At the end of the

[4] See Lützen (1990, 398) for a copy of this page.

paper, Liouville remarked that his analysis could "be generalized right away to the case where one joins to these three signs [algebraic functions, logarithms, and exponentials] the sign $\int$, indicating an indefinite integral... . In fact the functions which are created by the use of the sign $\int$ possess, in this type of research, properties completely analogous to those of the logarithms and can be treated by the same methods" (Liouville 1839, 456).

If one can express a function as a finite expression in terms of algebraic functions logarithms, exponentials, and (indefinite) integral signs, it is said to be expressed by quadrature.[5] So Liouville here suggested that his methods may be extended to the question of solutions of differential equations by quadrature.

His suggestion was substantiated and put to use in (Liouville 1840) where he showed that the elliptic functions of the first and second kind, considered as a function of their modules, are in general not expressible by quadrature and in (Liouville 1841) where he studied the famous Riccati equation

$$\frac{dy}{dx} + ay^2 = bx^m, \ (a, b, m \in \mathbb{R}).$$

Since the research of the Bernoullis in the 1720s, it had been known that the equation could be solved by quadrature for the following particular values of the modulus m:

$$m = -\frac{4n}{2n \pm 1} \ (n \in \mathbb{N}).$$

No other integrable cases had been found, so it was generally believed that the equation could only be integrated by quadrature for these values of m but Liouville was the first to provide a proof of impossibility in all other cases.

14.6 Later developments

Liouville's ideas had little immediate resonance but in the 1860s and 1870s P. T. Pépin and Lazarus Fuchs (1833–1902) complemented some of Liouville's investigations on solutions of second-order linear differential equations in algebraic form. Their research was carried further by Klein and Lie who used group theory. A more regular "Galois theory" of differential equations was developed by Leo Königsberger (1837–1921), Émile Picard (1856–1941),

[5] Though this expression originates in the ancient geometric concept of quadrature, the modern meaning has lost its geometric connotations. Here quadrature simply means the evaluation of indefinite integrals or antiderivatives.

and Ernest Vessiot (1865–1952) from 1870 on. For example Vessiot could in 1892 show how the structure of the Galois group of a linear differential equation could reveal whether the equation can be integrated by quadrature or in algebraic terms. The core of Liouville's theory, the question of deciding whether an integral can be written in finite form, was taken up in a few papers by the Russian mathematicians Mikhail Ostrogradsky (1801–62) and Pafnuti Lvovich Chebychev (1821–94) in the 1850s and the Englishman Godfrey Harold Hardy (1877–1947) at the beginning of the twentieth century. However, the classical account of the field only appeared in 1948 in the form of a book entitled *Integration in Finite Terms; Liouville's Theory of Elementary Methods* written by Joseph Fels Ritt (1893–1951). As indicated in the title, Ritt stayed close to Liouville's methods but clarified some of the function theoretical aspects that had been swept under the carpet by Liouville.

We saw how Liouville consciously strove to use primarily algebraic arguments in his publications on integration in finite terms. This tendency was carried to the limit by Maxwell Rosenlicht (1968), who defined what he called a differential field in which he could formulate many of Liouville's results abstractly.

The most spectacular advance in the theory of integration in finite terms came in 1970 when Robert Risch (b. 1939) announced that he had found a finite algorithm that can determine whether the integral of a given explicit finite function is again a finite function and find the integral when it is (Risch 1970). Since then, this or similar algorithms have been implemented on computers. That means that if a good computer program declares that it cannot find a simple expression of an integral, it probably means that it is impossible to write it in finite explicit terms.

14.7 Concluding remarks on the situation *c*.1830

It is conspicuous that several impossibility results were proved around 1830: The solution of the quintic was proved to be impossible by radicals, the two classical problems of the duplication of a cube and the trisection of an angle were proved to be impossible by ruler and compass, and it was proved that many integrals could not be expressed in finite terms. Why this sudden outburst of impossibility theorems? No doubt one reason is a technical mathematical one: The search for the solution of the quintic by radicals had developed algebra to a state where it could be used to prove such impossibilities. Indeed although the three mentioned impossibility results seem to deal with three quite different areas of mathematics, algebra, geometry, and integral calculus, it was the new algebraic methods and concepts such as permutations and the irreducibility of polynomials that provided the means for the impossibility proofs. Without these methods, all the earlier attempts to prove the impossibilities had been fruitless or at least not entirely convincing.

Moreover, the gradual change in the meta-mathematical view of the purpose and meaning of mathematics increased the importance of impossibility results in the mathematical community. In particular, the nineteenth-century experienced a gradual turn to a more conceptual view of mathematics where impossibility results could be considered as solutions of problems. As we saw earlier, Abel in particular voiced this new viewpoint.

Still, it was only a small number of young mathematicians who contributed to the proofs of impossibility: Abel, Wantzel, and Liouville. We have seen that Wantzel's results were hardly noticed at the time. Abel's and Liouville's achievements were recognized by the mathematical community, but very few mathematicians continued their research. To be sure, Galois continued Abel's research on the solvability of equations, but his ideas were not developed further until about 1870. As we shall see in Chapter 16, we have to wait until the end of the nineteenth century before the general mathematical community began to consider impossibility theorems to be as interesting and important as positive theorems.

15
Impossibility of Proving the Parallel Postulate

In the previous chapters, we have considered impossibility theorems establishing that a particular object or construction does not exist within a given mathematical theory. This chapter is devoted to the first impossibility result establishing that a proof of a statement does not exist. If a statement can be proven to be false in the sense that its negation can be proved, it is clear that one cannot prove the statement itself, at least if the theory is consistent. The surprising outcome of the story we are going to follow in this chapter is that an intuitively plausible statement could be shown to be impossible to prove at the same time as its negation cannot be proved either. Given the infinitely many possible deductions one can make from a collection of known theorems it is not at all obvious how one should be able to prove the non-existence of such a proof. The method for accomplishing such a task was discovered in the 1860s and it was so surprising that the mathematician who came upon it did not see what he had found until a colleague pointed it out to him.

The first statement that was proven to be unprovable is the so-called parallel postulate. The status of that postulate was debated for more than two millennia, resulting in the invention of a new type of geometry, the so-called non-Euclidean geometry or hyperbolic geometry. Its discovery or invention was instrumental in shaping our modern idea of physical space and our philosophical understanding of the nature of mathematical theories and in particular the nature of axioms at the basis of such theories. Thus, the development we shall follow in this chapter is one of the conceptually most important in the history of mathematics.[1]

[1] There are several good books on the history of non-Euclidean geometry including Bonola (1955), Rosenfeld (1988), Gray (1989), and Toretti (1984). Stillwell (1996) contains English translations of important papers on the subject.

A History of Mathematical Impossibility. Jesper Lützen, Oxford University Press.
 DOI: 10.1093/oso/9780192867391.003.0015

15.1 The axiomatic deductive method

Most sciences are characterized by the objects they deal with. Zoology deals with animals, astronomy deals with heavenly bodies, etc. However, mathematics is best characterized by its method. It is the science where we insist on *proving* all results using the most rigorous logical deductions. This insistence on proof was introduced by the ancient Greeks around 400 BC and has since penetrated mathematics all over the world. The Greeks also realized that one cannot prove everything. Indeed, as pointed out by Aristotle, in order to deduce a proposition one must base the argument on something known, something more fundamental. But in order to prove the more fundamental propositions one must rely on even more fundamental propositions and so on. In order to avoid an infinite regress we must therefore base our deductions on some fundamental propositions that we simply postulate to be true. They are the so-called axioms of the mathematical theory in question. For this reason, we call mathematics an axiomatic deductive science.

The first still extant mathematical treatise written according to the axiomatic deductive paradigm is Euclid's *Elements* from *c.*300 BC.[2] Euclid divided the axioms into two groups: the postulates and the common notions. There are five of each. In Chapter 4 we already listed the first three postulates, namely the so-called construction postulates. The two last Euclidean postulates state as follows:

4. That all right angles are equal to one another.
5. (The parallel postulate) That if a straight line falling on two straight lines makes interior angles on the same side less than two right angles, the two straight lines, if produced indefinitely, meet on that side on which are the angles less than two right angles.

With the notation of Figure 15.1 the parallel postulate states, that if the two angles u and v together are less than two right angles (or 180°) the lines l and m will intersect somewhere to the right if they are prolonged far enough.

The common notions are of a more logical and general nature than the postulates and we shall not deal with them. While Euclid based his deductions in the *Elements* on these explicitly formulated postulates and common notions

[2] See Euclid (1956). The postulates appear on pp. 154–5 of vol. 1 and the comments to postulate 5 on pp. 202–20.

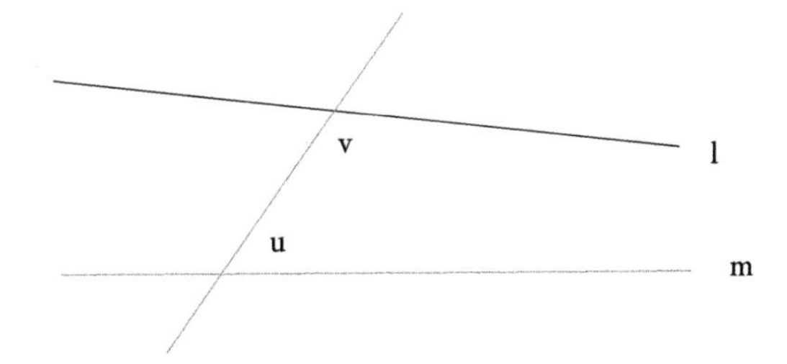

Figure 15.1 The parallel postulate: $u + v < 2R \Rightarrow l, m$ intersect

it is nevertheless the case that he sometimes used other basic "facts" that he either forgot to formulate explicitly as postulates or considered to be so evident that they need not be explicitly mentioned. In particular, he often drew implicit qualitative conclusions from an inspection of the drawing of the geometric figure. For example, he took it for granted that given a line segment, the two circles drawn with centers at the end points of the segment and both having the segment as radius will intersect each other. In the following, we will follow Euclid and his successors and assume these implicit axioms as well as the explicitly formulated postulates and common notions. A complete axiom system for geometry was not formulated until the end of the nineteenth century.[3]

How did the Greeks consider the philosophical nature of the postulates? While the common notions were considered as obvious truths, the postulates had a more subtle status. According to Aristotle, they were strictly speaking not true of the physical world. For example, the first postulate claims that one can draw a straight line through two given points. However, a straight line has by definition no width, but even the finest pen will draw a line of some finite width. Even worse, in the Aristotelian finite cosmology the parallel postulate is evidently false. If the angle sum $u + v$ in Figure 15.1 is very close to two right angles, the intersection between the two lines l and m may fall outside the sphere of the fixed stars so it will not exist in the physical Aristotelian world. Still, the postulates were considered true of an ideal or abstract world, which is described accurately by geometry.[4]

[3] Here I use the word "complete" informally, meaning that one can deduce the usual theorems of Euclidean geometry from the axiom system.

[4] According to de Risi (2016), Leibniz had a different idea of existence of geometric objects. He believed that it is possible to introduce a concept into geometry if its introduction does not lead to a contradiction.

15.2 The parallel postulate and the attempts to prove it

Even a superficial view of Euclid's five postulates reveals the special status of the last one, the parallel postulate. It is much longer and concerns a more complicated situation than the first four postulates. Already in antiquity, the parallel postulate was considered less evident than the other postulates. Not only is it not true of the Aristotelian universe, it is even not entirely obvious about an ideal infinite space. Without using the parallel postulate Euclid proved that if the angle sum $u + v$ in Figure 15.1 is equal to two right angles the two lines cannot intersect; in other words they are parallel. The parallel postulate then claims that if we rotate one of the two lines just a little bit, the two lines will intersect. If we only rotate very little it is obvious that the intersection point will be very far away. Could it not be possible that one could rotate by an angle so tiny that the lines will remain non-intersecting? Could it, for example, happen that the two lines would get closer to one other, just like a hyperbola and its asymptote, without ever intersecting?

Such questions were already discussed before Euclid. Still, Euclid decided to include the parallel postulate among his basic axioms. He had probably realized (without having a strict proof) that it (or another similar postulate) was needed in order to prove many fundamental theorems in geometry such as Pythagoras' theorem or the theorem stating that the angle sum in a triangle is equal to two right angles. Today we consider Euclid's inclusion of the parallel postulate among the basic axioms as a stroke of genius. However, it was controversial from the start and was often considered a blemish on the *Elements*. Euclid himself seems to have been aware of the special status of the parallel postulate. Indeed, he postponed the use of it as long as he could. It is used for the first time in proposition 29 of the first book of the *Elements*.

Most of Euclid's critics did not question the correctness of the postulate, but thought that it was too complex to play the role of a postulate. It ought to be deduced from other postulates and perhaps another more basic postulate. Of course, if it were possible to prove the parallel postulate from other postulates, it would be a theorem and could then be removed from the list of postulates. Many "proofs" were put forward by ancient Greeks, by medieval Arab mathematicians, and by European mathematicians from the Renaissance to the mid-nineteenth century. From a modern point of view, all the proofs contain flaws and, in fact, they were never accepted by a majority of the contemporary mathematical community.

Some of the "proofs" were combined with an alternative definition of parallel lines. Euclid defined parallel lines as lines in the same plane that do not intersect each other if produced indefinitely in both directions. Posidonius (*c.*100 BC), on the other hand, defined parallel lines as lines in the same plane having equal perpendicular distance everywhere. The change of definition of parallel lines sounds as an innocent change that should not affect the provability of the parallel postulate. After all, the parallel postulate does not mention parallel lines at all. The problem arises only because the mathematicians using the Posidonian definition also assume that the line (i.e., curve) that has a constant distance to a straight line is itself a straight line. Isn't that obvious? No! Just think of the corresponding situation on the surface of a sphere, e.g., the Earth: There, great circles (for example, the equator) play the role of straight lines (they are locally shortest lines between two points). But a curve having a constant distance to a great circle is a small circle which does not correspond to a straight line. However, if one assumes that a line equidistant to a straight line is again a straight line, one can prove the parallel postulate.

An example of such a proof was put forward by the Arab mathematician Ibn al-Haytham (965–1040). It is based on the well-established equivalence between the parallel postulate and the following theorems:

1. The angle sum of a triangle is equal to two right angles.
2. The angle sum of a quadrilateral is equal to four right angles.
3. In a quadrilateral with three right angles, the opposite sides are equal.

Ibn al-Haytham proved the latter:

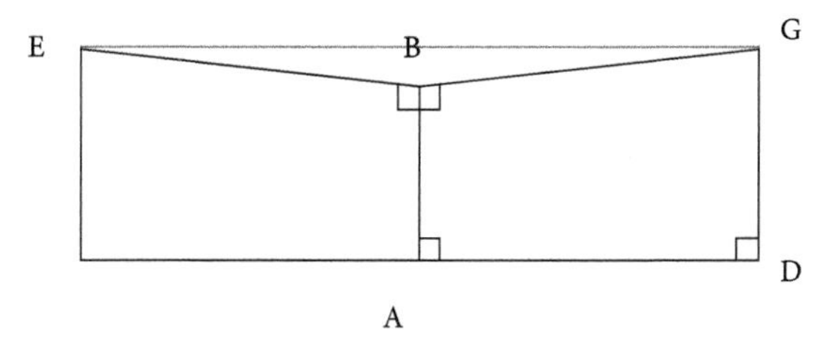

Figure 15.2 Ibn al-Haytham's proof of the parallel postulate

Let *ABGD* be a quadrangle with right angles at *A*, *B*, and *D* (Figure 15.2). Reflect the figure in *AB*. Move the line *DG* toward the left, remaining perpendicular to *AD* until *D* reaches its mirror image. Then it is clear that *G* has reached its mirror image *E*. If we assume that the topmost distance line *EG* traced out by *G* is a straight line, we can show that *AB* is equal to *DG*. Indeed the line *ABG* is straight, since the two angles at *B* are right angles. But if *AB* is either shorter or longer than DG, B will not lie on the top distance line. Thus, if the distance line is straight, there would be two different straight lines between the points E and G. This contradicts the usual reading of Euclid's first postulate according to which there is only one straight line between two points.

Thus, the assumption that a line equidistant to a straight line is straight implies the parallel postulate. We can therefore replace the parallel postulate by the mentioned assumption. However, it is not obvious that such a replacement leads to a simplification of the axiom system.

Another simple proof was suggested by John Wallis (1616–1703). He attacked Euclid's postulate straight on:

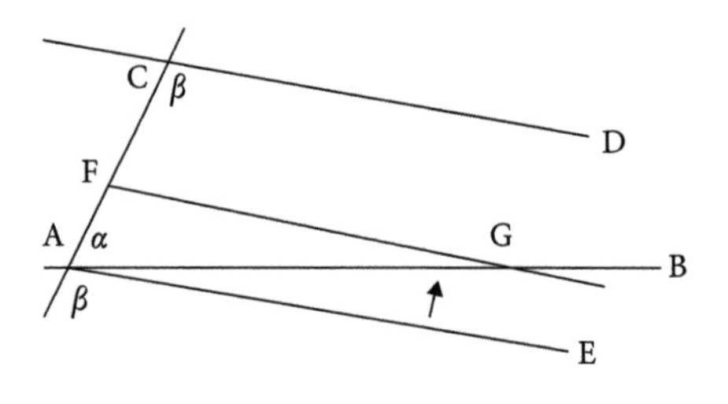

Figure 15.3 Wallis' proof of the parallel postulate

Let the two half-lines AB and CD (Figure 15.3) meet the transversal AC under the internal angles α and β and assume that $\alpha + \beta$ is less than two right angles. We must prove that the half lines intersect. Through A we draw a half line AE making the angle β with the transversal. Since $\alpha + \beta$ is less than two right angles it is clear that AE will lie under the half-line AB. Now move the half-line up a little bit so that its left end point will fall in F and such that the angle with the transversal remains equal to β. It is clear (by continuity) that we can move the half-line upward so little that some point on it, say E, remains under the half-line AB. In that case, there will on the moved half-line be a point F above AB and a point below AB. Thus, the two half-lines will intersect[5] at a point we can call G.

Now Wallis explicitly used a property that he considered more obvious than the parallel postulate: To any triangle there exists a similar triangle of any size. Apply this postulate to the triangle AFG and enlarge it so that the side AF is enlarged to the size AC. Since the angle AFG is equal to β the enlarged triangle will have the sides along AB and CD and they will therefore intersect in the corner corresponding to G in the enlarged triangle. QED.

It turns out that if we only know that there exist two similar figures of different size, we can prove the parallel postulate. While Wallis showed how to

[5] Here we use an axiom that is not explicitly stated by Euclid, but often used by him.

replace the parallel postulate by another seemingly more plausible postulate, other mathematicians continued to search for a proof based on the rest of Euclid's postulates alone.

15.3 Indirect proofs: Implicit non-Euclidean geometry

One of the most interesting attempts to prove the parallel postulate was published by the Italian Jesuit Girolamo Saccheri (1667–1733) in the year of his death.[6] His proof is a long but, in principle, elementary double indirect proof. He considered a figure that had also been essential for Omar Khayyam (1048–1131) and other Arab mathematicians: A quadrilateral *ABCD* with two right angles at *B* and *C* and the sides *AB* and *CD* equal to each other (Figure 15.4). He could easily prove that the two remaining angles at *A* and *D* are equal. Thus, there are three possible situations:

1. The hypothesis of the right angle where the angles at *A* and *D* are right angles.
2. The hypothesis of the obtuse angle where the angles at *A* and *D* are larger than a right angle.
3. The hypothesis of the acute angle where the angles at *A* and *D* are smaller than a right angle.

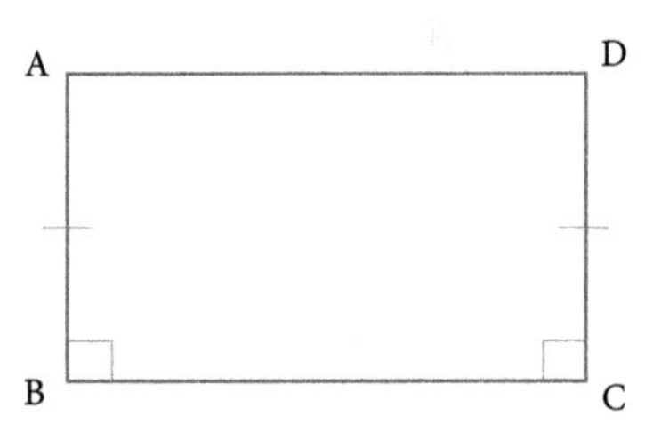

Figure 15.4 A Saccheri quadrilateral

Saccheri could easily prove that the hypothesis of the right angle was equivalent to the parallel postulate. He would therefore have proven that postulate, if he could prove that the other two hypotheses would lead to a contradiction.

First, he disposed of the hypothesis of the obtuse angle. Through a number of correct deductions, he could prove that that hypothesis would imply the hypothesis of the right angle. It would therefore "destroy itself" as Saccheri

[6] Saccheri's *Euclides ab omin naevo vindicatus* has been translated into English (Euclid freed of every fleck) by Bruce Halsted (Saccheri 1733).

put it. One may wonder how he could prove such a contradiction given that the hypothesis of the obtuse angle holds true on the sphere (with great circles playing the role of straight lines). The reason lies in the assumed infinity of the straight line that Saccheri and his fellow geometers took for granted. This assumption does not hold on a sphere, where a great circle will run back into itself if prolonged sufficiently far.

The assumption of the acute angle turned out to be more difficult to get rid of. However, after more than 50 pages of intricate but entirely correct deductions Saccheri arrived at a consequence that he claimed would "contradict the nature of the straight line." If he had been right, he would indeed have proved that the hypothesis of the right angle and consequently the parallel postulate would be deductible from the other postulates, and would therefore be superfluous as a separate postulate.

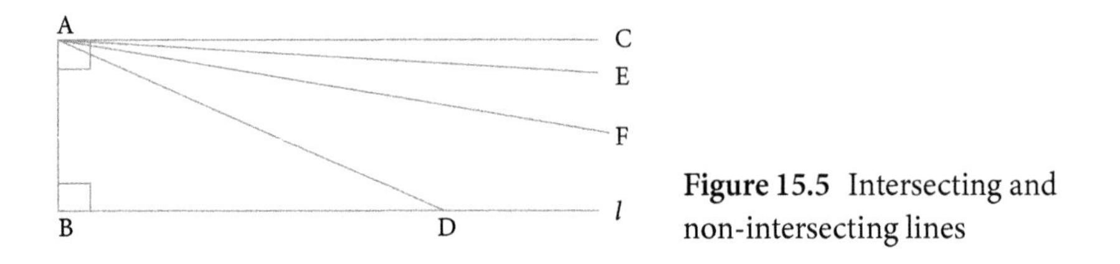

Figure 15.5 Intersecting and non-intersecting lines

However, his contemporaries and successors did not agree that he had, in fact, reached a contradiction. He had shown that under the hypothesis of the acute angle straight lines would behave in strange ways. Given a straight line *l* and a point *A* outside the line, there would be several lines through *A* that do not intersect *l* (Figure 15.5). In addition to the perpendicular *AC* to the perpendicular *AB* to *l* there would be non-intersecting lines like *AE* below *AC*. There would also be intersecting lines like *AD*, and a line *AF* separating the non-intersecting lines above it from the intersecting lines below it. *AF* would itself be non-intersecting, and thus the lowest non-intersecting line through *A*. There would be a similar situation to the left of *AB*.

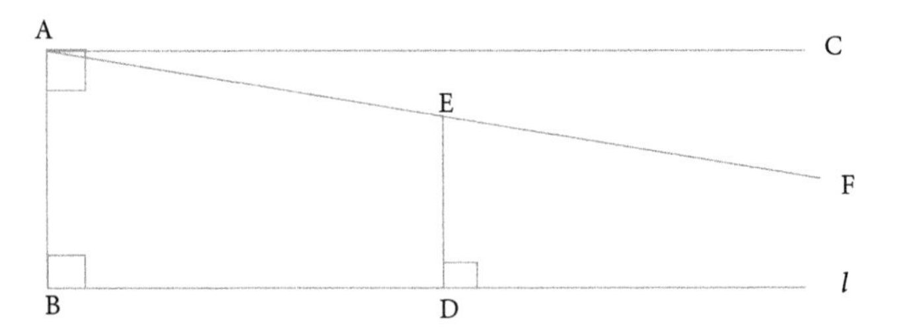

Figure 15.6 The limit parallel *AF* to *l*

Saccheri then derived several theorems about the separating limit parallel *AF* (Figure 15.6)[7]: As the point *D* on *l* moves toward the right, the perpendicular *DE* will decrease in length, and in fact it will tend to zero when *D* goes to infinity. The angle *DEF*, which Lobachevsky later called the parallel angle corresponding to the line segment *DE*, will always be less than a right angle, but it will increase as *D* moves to the right and will tend toward a right angle as *D* tends to infinity. From this, Saccheri concluded that *AF* will intersect *l* at a point at infinity and at that point, the two lines will have a common normal. However, he claimed that it is contrary to the nature of the straight line that two intersecting straight lines in the same plane could have a common normal at their point of intersection. Thus, he thought he had arrived at the desired contradiction.

To be sure, if the intersection point between two different straight lines is an ordinary point, the contradiction would have been manifest. However, if we invent new points at infinity, as is actually done in projective geometry, it is not certain that they will have the same properties as the usual points. And so the contradiction evaporates.

Although posterity did not accept Saccheri's "proof" of the parallel postulate, his essay contains interesting results. Indeed, until he arrived at his "contradiction" Saccheri was very careful not to use any property that build on the parallel postulate. The results he arrived at under the hypothesis of the acute angle, including the behavior of the limiting parallel, are therefore theorems in a geometry, in which the angle sum of a quadrilateral is smaller than four right angles or the angle sum in a triangle is smaller than two right angles. To be sure, for Saccheri the "results" were not real theorems but just steps toward the final contradiction. Still, after the invention of non-Euclidean geometry, Saccheri's steps would be regarded as theorems in this new geometry.

The Swiss mathematician Johann Heinrich Lambert (1728–77) continued in the footsteps of Saccheri. Like the latter, Lambert deduced a number of strange consequences of the hypothesis of the acute angle. Under that hypothesis, the angle sum of a triangle will be less than two right angles ($2R$) by an amount called the defect. Lambert realized that the defect is proportional to the area of the triangle:

$$Area\left(\Delta ABC\right) = K\left(2R - \left(\angle A + \angle B + \angle C\right)\right) = K \cdot defect\left(\Delta ABC\right).$$

[7] The derivation of these theorems is correct, but quite long. The theorems are true in the non-Euclidean geometry developed later by Lobachevsky and Bolyai.

In spherical geometry where the hypothesis of the obtuse angle holds, the angle sum of a triangle is greater than two right angles, and there the angle excess (i.e., the angle sum minus two right angles) is proportional to the area of the triangle. More precisely

$$Area\,(\Delta ABC) = r^2\left((\angle A + \angle B + \angle C) - 2R\right) = r^2 \cdot excess\,(\Delta ABC),$$

where r is the radius of the sphere. Lambert remarked that if in the last formula we replace r with ir where i is the imaginary unit, we will get the first formula for the defect with $K = r^2$. This made him propose that the hypothesis of the acute angle holds on the surface of an imaginary sphere. In a sense this was the first model of non-Euclidean geometry, but it lacked an intuitive geometric meaning.

In the end, Lambert seems to have given up refuting the hypothesis of the acute angle. This may explain why his paper, written in 1766, was only published posthumously in 1786.

Around 1800 Adrien Marie Legendre Legendre (1752–1833) published consecutive editions of a new *Elemens de géométrie* that was intended to replace Euclid's *Elements*. Legendre believed he could avoid the use of the parallel postulate by proving that the angle sum in a triangle is two right angles. However, despite the great influence of Legendre's book, his successive attempts to prove that the angle sum of a triangle cannot be smaller than two right angles build on various implicit assumptions that turned out to be equivalent to the parallel postulate. As all other proofs, they were rejected by most of his contemporaries and successors.

15.4 Non-Euclidean geometry: The invention

During the 1820s three mathematicians Carl Friedrich Gauss (1777–1855), Nikolai Ivanovich Lobachevsky (Figure 15.7), and Janos Bolyai (1802–60) independently set out to develop what Gauss called a non-Euclidean geometry.[8] They had all tried to prove the parallel postulate and had all become convinced that it couldn't be done. If the parallel postulate cannot be proved, it is logically possible to develop a new geometry satisfying the negation of the parallel postulate or equivalently Saccheri's and Lambert's hypotheses of the acute angle. And that was what they did.

[8] Translations of Lobachevsky's and Bolyai's treatment of non-Euclidean geometry can be found in Bonola (1955).

Figure 15.7 Nikolai Ivanovich Lobachevsky (1792–1856)
Reproduced by Jim Høyer DMS

The new geometry had many strange properties: The angle sum of a triangle is smaller than two right angles and the defect is proportional to the area of the triangle. That means that there is an upper limit to the size of triangles and there do not exist similar triangles of different size. This is, of course, an architect's nightmare. It is impossible to reduce the size of a plan of a house without distorting the design.

Moreover, there would exist an absolute measure of length, i.e., a unit of length that one would be able to convey to another person over the phone without reference to a physically existing measure stick. In ordinary Euclidean geometry there exists an absolute measure of angle. Indeed the Euclidean definition of a right angle is such an absolute definition: Intersect two straight lines. If two angles at the point of intersection next to each other are equal, they are by definition right angles. However, in Euclidean geometry there is no way one can explain the length of a meter without referring to a meter stick or another physical phenomenon, the length of which you specify in meters. In the new non-Euclidean geometry there is a connection between angles and lengths that makes it possible to define an absolute measure of length. For example, Gauss suggested that one could choose as unit of length, the side of an equilateral triangle with the angles equal to 59°59′59″.99999. Since there is precisely one such triangle in non-Euclidean geometry, this definition determines the unit of length uniquely and absolutely.

In the new geometry, there would be, as in Figure 15.5, many lines through a given point A that do not intersect a given line l that does not pass through the given point. The boundary line AF separating the intersecting and the non-intersecting lines is often called the parallel to l. Its distance to l will tend to zero as we go to the right in the figure. If we consider a given line l and its parallels through different points outside l we will have a family of parallels as in Figure 15.8.

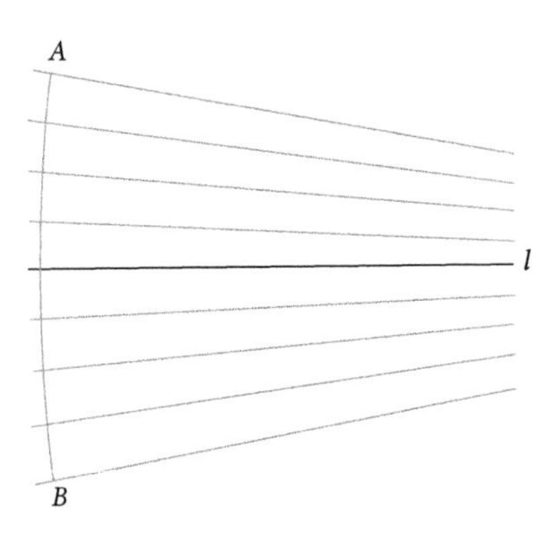

Figure 15.8 Parallel lines and horocycle AB

The curve AB that intersects all the parallel lines orthogonally is called the horocycle. If it is rotated around one of the parallels the result is a surface called the horosphere. It is in a sense a sphere with infinite radius. In Euclidean geometry this surface would be a plane. In non-Euclidean geometry, Gauss, Lobachevsky, and Bolyai all discovered that it is not a plane but a surface on which the geometry is in a sense Euclidean. To be sure, it is not straight lines but horocycles that lie on the horosphere, but we can consider horocycles as "lines" (just as we consider great circles as lines in spherical geometry). If we do that, it turns out that a triangle on the horosphere has an angle sum equal to two right angles, and the parallel postulate holds on the horosphere. From this interesting fact, Lobachevsky and Bolyai developed a non-Euclidean trigonometry which differs from the Euclidean one.

In non-Euclidean three-dimensional geometry one can, of course, consider a sphere and the spherical geometry on it. The three inventors of non-Euclidean geometry discovered that this spherical geometry was exactly the same as spherical geometry on a sphere in Euclidean geometry.

Although many of the results of the new geometry seem strange and counterintuitive, the three inventors of non-Euclidean geometry did not consider them paradoxical. Indeed, Gauss, Lobachevsky, and Bolyai considered non-Euclidean geometry as a possible description of physical space. For them and for many of their successors such as Riemann, Helmholtz, and Clifford, the investigation of the status of the parallel postulate was not just a purely mathematical and logical exercise concerning consistence and completeness of an axiom system. It was an essential part of an investigation of the nature of the space we inhabit.

Gauss's, Lobachevsky's, and Bolyai's research on non-Euclidean geometry remained virtually unnoticed until the late 1860s. There are many reasons for that.

1. Gauss was the most celebrated mathematician of his generation but deliberately kept his ideas on non-Euclidean geometry secret. He published nothing on the matter and demanded his correspondents to keep quiet about his unconventional ideas. He was afraid of the reaction of the "Boetians."
2. Lobachevsky and Bolyai were rather unknown mathematicians working in Kazan and Hungary, respectively, far away from the mathematical centers of the time. Bolyai published his ideas in 1833 as an appendix (in Latin) to his father's not so widely circulated geometry book. Most of Lobachevsky's books and papers were published in Russian, but he did publish a few papers in French and German, including one in1840 in *Crelle's Journal*, the most influential German mathematical journal at the time. Their main papers can be found in translation in (Bonola 1955).
3. The ideas were controversial and in direct opposition to the ideas of Immanuel Kant (1721–1804), the most influential philosopher at the time. According to Kant, space was an a priori intuition, necessary for ordering our perceptions. And for Kant, Euclidean geometry was the only possible geometry of space. In fact, he used Euclid's proof of the angle sum in a triangle to argue that the properties of space could be obtained by constructions in our intuition without any appeal to empirical perceptions. Gauss and Lobachevsky argued that Kant was wrong. According to them, the nature of space was an empirical question.
4. In fact, they both referred to surveying and astronomical data that could determine the question, but they found no perceptible deviation from Euclidean geometry.
5. Finally, and from a logical standpoint most importantly, the three inventors of non-Euclidean geometry had no convincing argument demonstrating that their new geometry was consistent. To be sure, they had not found any contradictions in their new theory. But who says that

further deductions would not reveal a contradiction, such as Saccheri had assumed? Lobachevsky argued, that since his new geometry led to correct analytical formulas, it must represent a viable type of geometry. However, his argument is not valid. Indeed a contradictory theory can imply correct results.

15.5 The help from differential geometry of surfaces

The solution to the last-mentioned problem of consistency came from another part of geometry, the differential geometry of surfaces. Here Gauss in 1828 began a new approach called intrinsic geometry.[9] It is that part of the geometry of a surface that is not dependent on how the surface is embedded in three-dimensional space. Said differently, with a metaphor invented by Helmholtz, it is the geometry that a two-dimensional surface dweller in the surface would experience. For example, the surface dweller would consider locally shortest lines, so-called geodesics, to be straight. Gauss introduced a new concept of the curvature of a surface, now called its Gaussian curvature. He could prove that it is an intrinsic property of the surface in the sense that it does not change if the surface is bent without stretching. This is his celebrated Theorema Egregium (eminent theorem). Thus, the Gaussian curvature is a property that a surface dweller can calculate from intrinsic measurements in the surface. The Gaussian curvature is positive in points where the surface looks like a sphere and negative in points where the surface looks like a saddle, bending one way in one direction and another way in the orthogonal direction.

Gauss also proved a theorem, later called the Gauss–Bonnet theorem, relating the curvature inside a geodesic triangle with its angle sum. In the special case where the Gauss curvature is constant and negative, the theorem states that the angle defect is proportional to the area of the triangle. This was precisely the same theorem he had shown to hold in non-Euclidean geometry. Thus, a surface dweller on a constant negatively curved surface would experience living in a non-Euclidean universe.

It is not clear whether Gauss himself made this connection between his theory of surfaces and his secret non-Euclidean geometry. The first one to make the connection explicit was Eugenio Beltrami (Figure 15.9) in an 1868 paper entitled "Essay on the Interpretation of Non-Euclidean Geometry" (Beltrami 1868). He proved that on a surface of constant negative curvature that he called

[9] Gauss' *Disquisitiones circa superficies curvas* was translated into English in 1965 (Gauss 1828).

a pseudosphere, all of Lobachevsky's theorems of non-Euclidean geometry would hold. He thus considered such a surface to be a "real substrate" for non-Euclidean geometry. Today we would call it a model. At first, he remarked that he still had not given up proving the parallel postulate, but soon Guillaume-Jules Houël (1823–86), who translated his paper into French, made it clear to him that his model, in fact, proved that the parallel postulate cannot be proved.

Figure 15.9 Eugenio Beltrami (1835–1900)
Source: Archives of the Mathematisches Forschungsinstitut Oberwolfach

In order to see that, assume that one could prove the parallel postulate from the other postulates of Euclidean geometry. In that case non-Euclidean geometry would contain a contradiction, because the negation of the parallel postulate is assumed to hold. But that would be translated back via the model to a contradiction in the geometry on a surface in Euclidean geometry. There would thus be a contradiction in Euclidean geometry. Thus, if Euclidean geometry is consistent, as all mathematicians at the end of the nineteenth century believed, there could be no contradiction in non-Euclidean geometry either. Thus, the parallel postulate cannot be proved from the other postulates.

Beltrami realized that his model had some problematic features: First, he believed (and Hilbert later proved) that one cannot find a surface of constant negative curvature without singularities imbedded in Euclidean three-dimensional space. The most famous pseudo-sphere is the trumpet-shaped surface obtained by rotating a curve called the tractrix around its generator

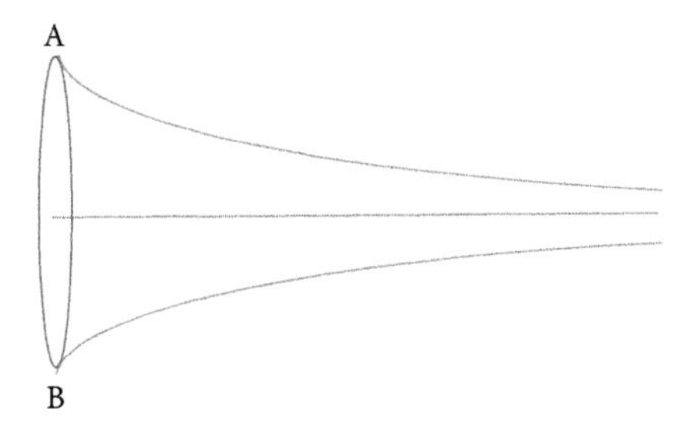

Figure 15.10 A pseudo-sphere obtained by rotating a tractrix

(see Figure 15.10). It has singularities along the circle AB where the tangent of the tractrix is vertical. Thus, like any other imbedded surface of constant negative curvature it cannot be extended to represent the entire non-Euclidean plane. Beltrami overcame this problem by building a model or a map of the entire non-Euclidean plane on a Euclidean circular disk equipped with a special metric. A metric is a method for determining distances. Beltrami's metric was built in such a way that two points with a constant Euclidean distance have a Beltrami distance that increases toward infinity as the point pair moves from the center of the disk toward its periphery.

Another problem was that Gauss's surface theory only allowed Beltrami to make a model of non-Euclidean plane geometry. At first, he argued that it would be impossible to extend the idea of a pseudo-sphere to three-dimensional space. However, his proof only established the rather obvious fact that one cannot build a model of three-dimentional non-Euclidean geometry if the metric is assumed to be inherited from the Euclidean metric in three-dimensional space. Yet, Riemann's revolutionary generalizations of Gauss's intrinsic surface theory soon showed to Beltrami how he could generalize the disc model to a sphere model of three-dimensional non-Euclidean geometry.

Bernhard Riemann (1826–66) wrote his paper "On the Hypotheses Which Lie at the Foundation of Geometry" for a talk he delivered in 1854 as part of his doctoral habilitation. It was published posthumously in 1868, the same year as Beltrami's first paper on non-Euclidean geometry. Riemann's paper was packed with new ideas but had almost no calculations to substantiate them. It presented a general notion of an n-dimentional manifold and a generalization of Gauss's metric on surfaces to such higher dimensional manifolds. Riemann insisted that the metric properties of space could only be determined empirically. Where Lobachevsky and Bolyai had studied geometries corresponding to constant curvature, Riemann entertained the possibility that the curvature of physical space could vary from point to point as long as the average curvature over measurable distances was close to zero. The British mathematician William Kingdon Clifford (1845–79) in 1876 extended Riemann's ideas in an attempt to explain all of physics, matter in particular, as epiphenomena of the geometry of a space of changing curvature. He did not get far with this idea, but in 1916 Albert Einstein (1879–1965) built on Riemann's ideas and its developments in so-called tensor calculus when he discovered his general theory of relativity which is still our best model of space-time.

The renaissance of non-Euclidean geometry around 1870 was triggered by the publication beginning in 1866 of Gauss's correspondence on non-Euclidean geometry and soon thereafter by the publication in 1868 of the works of Beltrami, Riemann, and Hermann von Helmholtz (1821–94). From then on, the new geometry was widely known although not uniformly approved. In particular, philosophers had a hard time accepting the revolutionary new ideas. Where mathematicians soon argued that non-Euclidean geometry was a priori just as true as Euclidean geometry, because the two geometries were equally consistent, some philosophers interpreted Beltrami's discovery as the deathblow to non-Euclidean geometry. Indeed, they argued, Lobachevsky's and Bolyai's ideas did not describe a new geometry after all, but just the well-known geometry on a surface of constant negative curvature in Euclidean space. Thus, they concluded, Euclidean geometry is the only geometry as Kant had already argued.[10]

15.6. Conclusions

The invention of non-Euclidean geometry had wide-ranging consequences. In physics, it opened the possibility that space could be described by a geometry different from the one Euclid and his followers had imagined for two millennia. The best mathematical model of physical space could no longer be determined a priori by philosophers or mathematicians but had to be determined by experience and convenience. At first, most physicists considered this freedom a purely theoretical possibility but with Einstein, new geometries entered mainstream physics.

In mathematics, the proof that non-Euclidean geometry was just as consistent as Euclidean geometry began a re-evaluation of the role of axioms in mathematics. Where they had previously been considered as more or less evident truths about some possibly idealized part of the real world, they now

[10] For highly interesting accounts of this renaissance of non-Euclidean geometry see Voelke (2005) and Volkert (2013).

became, in principle, arbitrary starting points for our deductions. We can choose to include the parallel postulate among our axioms or we can choose to include the negation of the postulate. Both possibilities are logically equally valid. Some axiom systems may be more convenient for modeling parts of reality, but if an axiom system is consistent, the theory it describes exists as a mathematical theory. We shall return to this formalist conception of mathematics in Chapter 17.

For the purpose of this book, the history of non-Euclidean geometry is interesting because it shows how it was discovered that one can prove that a particular axiom *A* is independent of the other axioms in an axiom system, i.e., that the axiom cannot be deduced from the remaining axioms of the system. The trick is to build a model satisfying the remaining axioms and the negation of axiom *A*. If the model is built within a mathematical theory *T* that we consider to be consistent, we can conclude that axiom *A* cannot be deduced from the other axioms. Indeed if *A* could be deduced from the other axioms both *A* and its negation would hold true in the model and thus in the mathematical theory *T* in which we have built the model. But that is not possible if that theory *T* is consistent. The method of modeling also opened a way to prove consistency of an axiomatically described theory *T*. One simply builds a model of it in another theory *T'*. If *T'* is consistent, *T* must be consistent as well. Indeed an inconsistency in *T* will turn up in the model as an inconsistency in *T'*. We say that the theory *T* is consistent relative to the theory *T'*.

For a modern mathematician this method for proving relative consistency and unprovability is so well known that it may seem almost trivial. Nevertheless, as we have seen in this chapter, Beltrami, the first person to model non-Euclidean geometry within Euclidean geometry, at first did not realize what he had established. Only after Houël had pointed it out to him did he realize that he had proved the unprovability of the parallel postulate. To be sure, this conclusion requires that we know that Euclidean geometry is consistent. Most mathematicians of the nineteenth century considered that to be obvious, so the requirement was not pointed out explicitly until Henri Poincaré (1854–1912) did so at the turn of the twentieth century.

16
Hilbert and Impossible Problems

16.1 Impossibility as a solution; rejection of *ignorabimus*

For more than a century, mathematicians from all over the world have gathered every four years for the International Congress of Mathematicians. The first of these congresses took place in 1897 in Zürich. On this occasion, the leading French mathematician Henri Poincaré gave a plenary address. Therefore, it was natural to ask his leading German colleague, David Hilbert (Figure 16.1), to give a talk in 1900 at the second International Congress of Mathematicians in Paris.[1] Following a suggestion by his friend Hermann Minkowski (1864–1909), Hilbert decided to talk about the course of mathematics in the new century.[2]

> Who among us would not be glad to lift the veil behind which the future lies hidden; to cast a glance at the next advances of our science and the secrets of its development during future centuries?
>
> (Hilbert 1900 in Gray 2000, 240)

In particular, he decided to talk about unsolved mathematical problems, the solution to which he believed could advance mathematics in the century to come. Problems, he declared, are the lifeblood of science:

> As long as a branch of science offers an abundance of problems, so long is it alive: a lack of problems foreshadows extinction or the cessation of independent development.
>
> (Hilbert 1900 in Gray 2000, 241)

As for the future development of mathematics, Hilbert was optimistic. In particular, he believed that all well-posed mathematical problems could be solved "for in mathematics there is no *ignorabimus*" (Hilbert 1900 in Gray

[1] In addition to this exception to the 4-year period of the International Congresses, the world wars also broke the rhythm.
[2] The quotations from Hilbert's lecture are taken from Gray's translation (Gray 2000).

A History of Mathematical Impossibility. Jesper Lützen, Oxford University Press.
 DOI: 10.1093/oso/9780192867391.003.0016

Figure 16.1 David Hilbert (1862–1943)
Reproduced by Jim Høyer DMS

2000, 248). With this assertion, he referred to a famous talk in 1872 by the German physiologist Emil du Bois-Reymond (1818–96). The subject of the talk was "The Limits of Science." Du Bois-Reymond had argued that in science there are fundamental things that we do not know and that are, in principle, unknowable (ignoramus et ignorabimus). Among such riddles, he mentioned the nature of matter and energy as well as the origin of motion. According to Hilbert, there is no reason for such pessimism in mathematics.

To be sure, Hilbert knew that in mathematics, there are impossible problems, but in these cases, he believed it would be possible to prove the impossibility. Here is what he had to say about impossibility theorems and in particular their recent acquisition of importance:

> Occasionally it happens that we seek the solution under insufficient hypotheses or in an incorrect sense and for this reason do not succeed. The problem then arises: to show the impossibility of the solution under the given hypotheses, or in the sense contemplated. Such proofs of impossibility were effected by the ancients; for instance, when they showed that the ratio of the hypotenuse to the side of an isosceles right triangle is irrational. In more recent[3] mathematics, the question of the impossibility of certain solutions plays a prominent part; and we perceive in this way that old and difficult

[3] "In der neueren Mathematik" is usually (e.g., by Gray 2000) translated into "In later mathematics." But the German phrase clearly does not just refer to a period after the ancients, but to Hilbert's recent past.

> problems, such as the proof of the axiom of parallels, the squaring of the circle, or the solution of equations of the fifth degree by radicals, have finally found fully satisfactory solutions, although in another sense than that originally intended. It is probably this remarkable fact along with other philosophical reasons that gives rise to the conviction (which every mathematician shares, but which no one has yet supported by a proof) that every definite mathematical problem must necessarily be susceptible of an exact settlement, either in the form of an actual answer to the question asked, or by the proof of the impossibility of its solution and therewith the necessary failure of all attempts.
>
> (Hilbert 1900 in Gray 2000, 247)

There are many remarkable aspects of this quote. First, Hilbert emphasized how impossibility theorems had acquired a prominent place in recent times. The examples he mentioned suggests that he thought of the nineteenth century in particular. Indeed, we have seen that prior to 1800 impossibility statements had often been considered less important than positive results. We have also seen that around 1830 several young mathematicians like Abel, Galois, Wantzel, and Liouville began to elevate impossibility theorems to a more important place in mathematics. Now Hilbert claimed that with these advances and the more recent proof of the transcendence of π and the proof of the impossibility of proving the parallel postulate, impossibility theorems had come to play a prominent (hervorragende) role. They had finally obtained full citizenship in the realm of mathematics.

Second, the recent proofs of important impossibility results had been instrumental in convincing Hilbert of what he in the next section called "the axiom of the solvability of every problem." This rejection of "ignorabimus" became an important part of his axiomatic approach to the foundation of mathematics to which we shall return in Chapter 17. Of course, the belief in the solvability of every mathematical problem presupposes that one accepts an impossibility proof as a solution of the problem although, as Hilbert put it, in a different sense than originally intended. Abel had already suggested that such a move would make it possible to answer every problem and with Hilbert this optimistic point of view was elevated to an axiom.

> This conviction of the solvability of every mathematical problem is a powerful incentive to the worker. We hear within us the perpetual call: there is the problem. Seek its solution. You can find it by pure reason, for in mathematics there is no *ignorabimus.*
>
> (Hilbert 1900 in Gray 2000, 248)

16.2 Hilbert's third problem: Equidecomposability

After a general introduction, Hilbert continued his 1900 lecture with a commented list of specific problems. In the written version, the list included 23 problems but in the oral presentation, he had only time to present 10 of them. The problems were quite different in nature: Some were very precise; others were broader and programmatic. Some of them were explicitly formulated as impossibilities; others were formulated in a positive way but eventually turned out to be impossible.

Hilbert's third problem is the first on Hilbert's list of problems that he formulated as an impossibility problem. The problem concerns the difference between the theory of areas of polygons and the theory of volumes of polyhedra. The theory of areas of polygons can be handled by finitary methods. To be more precise, if two polygons have the same area they can be divided into a finite number of pairwise congruent triangle polygons. Or said differently, it is possible to cut one of the given polygons into a finite number of triangles and then assemble them in a different way to obtain the other given polygon. If this can be done, we say that the two polygons are equidecomposable. In other words, two polygons with the same area are equidecomposable. This insight was proved at the beginning of the nineteenth century by William Wallace (1768–1843), Farkas Bolyai (1775–1856), and Paul Gerwien (*c.*1800–58?) and had been anticipated in Euclid's treatment of polygons. For example, Euclid established that two triangles with the same base and the same altitude have the same area by showing that they are equidecomposable.

Similarly, in dimension 3, two polyhedra are said to be equidecomposable if they can be divided into pairwise congruent tetrahedra. And analogously, Euclid proved that two tetrahedra with equal base and altitude have the same volume. However, he did not establish this equality of the volumes by showing that they were equidecomposable. Instead, his proof used the so-called method of exhaustion, which is a precursor of the modern method of integration. In a sense, this corresponds to cutting the tetrahedra into many infinitely small congruent parts.

In the early nineteenth century, Gauss emphasized this difference between dimension 2 and 3 in a letter to his colleague Gerling. Referring to this letter, Hilbert suggested that it would be impossible to deal with volumes using only finitary methods:

> Gerling also succeeded in proving the equality of volume of symmetrical polyhedra by dividing them into congruent parts. Nevertheless, it seems to

> me probable that a general proof of that kind for the theorem of Euclid just mentioned is impossible and it should be our task to give a rigorous proof of its impossibility. This would be obtained as soon as we succeeded *in specifying two tetrahedra of equal bases and equal altitudes which can in no way be split up into congruent tetrahedra, and which cannot be combined with congruent tetrahedra to form two polyhedra which themselves could be split up into congruent polyhedra.*
>
> (Hilbert 1900 in Gray 2000, 252)[4]

This problem was the first of the Hilbert problems to be solved. Already before Hilbert's lecture was published, his student Max Dehn (1878–1952) proved in 1901 that a regular tetrahedron and a cube of equal volume cannot be divided into congruent tetrahedra. Dehn's proof rested on the notion of invariants, i.e., quantities that do not change when a polyhedron is cut up into smaller pieces and then assembled to make up another polyhedron. The volume of the polyhedron is obviously an invariant and Dehn discovered another invariant. In order to form the Dehn invariant, consider an edge of a polyhedron and multiply its length (measured in some way) by the dihedral angle between the two surfaces meeting along the edge. Then sum these multiples over all the edges. If one calculates in a suitable way with these sums of products, it is not difficult to see that the Dehn invariant is indeed an invariant. Moreover, Dehn could show that it had different values for a regular tetrahedron and a cube of equal volume. Thus, these two polyhedra are not equidecomposable even if they have the same volume.

Thus, as Hilbert and other mathematicians had conjectured, the theory of volumes of polyhedra necessarily depends on infinite processes (or Archimedes' axiom, as Hilbert put it), whereas the theory of areas of polygons is independent of this axiom.

In his lecture, Hilbert emphasized that the solution of one problem often leads to new problems. This also happened in the case of his third problem. Indeed, Dehn had shown that if two polyhedra are equidecomposable, they *necessarily* have the same Dehn invariant. However, the equality of a Dehn invariant is not in itself sufficient to ensure that two polyhedra are equidecomposable. Thus, the question arose as to whether a necessary and sufficient condition could be found. In 1903, the Royal Danish Academy of Sciences and Letters announced a prize competition to solve this question. The academy

[4] The last property mentioned by Hilbert is called equicomplementability. I shall skip over the distinction between equidecomposability and equicomplementability.

only received one worthless contribution before the expiry of the deadline the following year. However, more than half a century later the academy decided to award the formally expired prize to the Swiss mathematician Jean-Pierre Sydler (1921–88), who had shown that two polyhedra are equidecomposable if and only if they have the same volume and the same Dehn invariant.

16.3 Hilbert's seventh problem

Hilbert's seventh problem deals with the "irrationality and transcendence of certain numbers." Such problems are, in fact, impossibility questions. Indeed, in order to prove that a given number is irrational one has to show that it is impossible to find two integers whose ratio is equal to the given number. In order to show that a given number is transcendental one has to show that it is impossible to find a polynomial equation with integer coefficients having the given number as a root.

We have noticed that Hilbert in the introduction to his lecture emphasized the solution of the classical problem of the quadrature of a circle as a remarkable recent impossibility theorem. As we saw in Chapter 13 the impossibility of a solution of the problem by ruler and compass was a consequence of Lindemann's proof of the transcendence of π. We also saw that Hilbert was greatly impressed by this and similar results and even published a new and shorter proof of the transcendence of π. In his seventh Paris problem he now suggested more general problems of this kind as worthy of mathematician's attention:

> Hermite's arithmetical theorems on the exponential function and their extension by Lindemann are certain of the admiration of all generations of mathematicians. But the task at once presents itself of penetrating further along the path here entered.
>
> (Hilbert 1900 in Gray 2000, 258–9)

In particular Hilbert formulated the following specific problem:

> The expression α^{β}, for an algebraic base α and an irrational algebraic exponent β, e.g., the number $2^{\sqrt{2}}$ or $e^{\pi} = i^{-2i}$, always represents a transcendental or at least an irrational number.[5]
>
> (Hilbert 1900 in Gray 2000, 259)

[5] Of course the base α must be different from 0 and 1.

Hilbert considered this problem as particularly difficult, more so than Fermat's last theorem. As it turned out, however, it was solved independently and affirmatively in 1934 by the Soviet mathematician Aleksander Gelfond (1906–68) and the German mathematician Theodor Schneider (1911–88) after the former had proven a special case in 1929. It gave rise to many striking results concerning transcendental numbers.[6]

16.4 Hilbert's first problem

Hilbert's first problem asks for the proof of two positive theorems about infinite sets. We shall discuss them here because the answers turned out to establish the impossibility of proving the theorems.

Since Greek antiquity, philosophers and mathematicians have debated the nature of the infinite. Aristotle distinguished between two types of infinity: actual and potential infinity. For example, the natural numbers are potentially infinite because given any finite set of natural numbers we can always find one more. Aristotle accepted the use of the potentially infinite. However, he argued that it would lead to paradoxes if one considered the "actual" infinite set of all natural numbers. The ban on actual infinities was followed by most antique mathematicians. For example, when Euclid proved the theorem that we today formulate thus: "There are infinitely many prime numbers," he wrote: "Prime numbers are more than any assigned multitude of prime numbers." That is precisely the potential infinity of the prime numbers. Similarly, in geometry, Euclid did not consider the entire straight line, but only a finite line segment. However, according to the second postulate it is potentially infinite, in the sense that it can be prolonged indefinitely: "To produce a finite straight line continuously in a straight line."

The distinction between actual and potential infinity and the exclusion of the former was accepted by the majority of mathematicians well into the nineteenth century. For example, Gauss vehemently rejected the use of actual infinite sets, and declared that the use of infinity in mathematics was just a manner of speaking. The paradoxes concerning actual infinities arose because of the general acceptance of Euclid's common notion 5: "The whole is greater than the part." Consider, for example, with Galileo Galilei (1638, 31–3) the set of square numbers {1, 4, 9, 16, ...}. To each natural number n, there corresponds precisely one square number, namely n^2.

[6] See, e.g., the chapter written by Gelfond himself in Alexandrov (1971).

Thus, there is a bijection between the set of all the natural numbers and the set of square numbers. In this sense, there are as many square numbers as there are natural numbers; and yet the set of square numbers is only a (small) part of all natural numbers. This contradicts Euclid's fifth common notion.

In the 1870s Georg Cantor (Figure 16.2) began to develop a theory of actually infinite sets. He said that two sets have the same cardinality (size) if there is a bijection between them. As pointed out by Galileo and others, he then had to accept that an infinite set could have the same cardinality as a proper subset. Cantor could show that natural numbers have the same cardinality as rational numbers. They are said to be countable. On the other hand, he could prove that the set of real numbers has a higher cardinality than the set of natural numbers in the sense that there does not exist a bijection from real numbers to natural numbers.

Figure 16.2 Georg Cantor (1845–1918)
Source: Archives of the Mathematisches Forschungsinstitut Oberwolfach

His second proof of this fact uses a famous diagonal argument: Assume that the real numbers between 0 and 1 are countable, so that we can write them all up in a numbered list $a_1, a_2, a_3, \ldots, a_n, \ldots$. Write the nth real number a_n in decimal form as $0, a_{n,1}a_{n,2}a_{n,3}a_{n,4}\ldots a_{n,n}\ldots$, where $a_{i,j}$ is a natural number between 0 and 9. Now consider a number $b = 0, b_1b_2b_3b_4\ldots b_n\ldots$, whose nth decimal b_n is different from $a_{n,n}$ (the diagonal element). Then b is different from a_1 because its first decimal place is different from that of a_1, it is different from a_2 because its second decimal place is different from that of a_2, etc., it is different from a_n

because its nth decimal place is different from that of a_n.[7] Thus, we have found a real number between 0 and 1 which is not in the numbered list, contrary to the assumption. Therefore, we have reached a contradiction and can conclude that the real numbers between 0 and 1, and thus a fortiori all the real numbers, cannot be counted.

In this way, Cantor had proved that there exist at least two essentially different infinite sets of numbers in the sense that they have different cardinalities. He asked himself if there could be other cardinalities in between the countable and the cardinality of the real numbers. He conjectured that the answer would be no. In other words, he believed that every infinite set of real numbers is either countable or of the same cardinality as the real numbers. The latter he called the cardinality of the continuum and for that reason, his conjecture is called the continuum hypothesis. Several times Cantor believed that he had proved the continuum hypothesis but every time he discovered that, his proof was incorrect.

Cantor's theory of infinite sets divided mathematicians. Some, like the German mathematician Kronecker, were strongly opposed to Cantor's actual infinite sets, whereas others like Hilbert enthusiastically embraced the new theory. In a lecture from 1925 Hilbert famously declared: "From the paradise that Cantor created for us, no one shall be able to expel us." Already in his 1900 Paris lecture on mathematical problems, he expressed his high regard for Cantor's theory by making it the subject of the first problem. He actually formulated two problems. The first one called for a proof of the continuum hypothesis. The other one asked for a proof of the so-called well-ordering theorem. It states that any set can be well ordered. Here a well-ordered set is an ordered set with the property that any non-empty subset has a first element. The usual ordering of the natural numbers is a well-ordering. On the other hand, the usual ordering of the real numbers is not a well-ordering. For example, there is no smallest positive number. Cantor believed that the well-ordering theorem was correct but found no convincing proof. In particular, he believed that it would be possible to equip real numbers with a well-ordering. Hilbert said in his 1900 address:

> It appears to me, to be most desirable to obtain a direct proof of this remarkable statement of Cantor's, perhaps by actually giving an arrangement of numbers such that in every partial system a first number can be pointed out.
> (Hilbert 1900 in Gray 2000, 250)

[7] I have left out a technicality, arising because some rational numbers can be written in two ways in decimal form. Moreover, Cantor actually expressed the real numbers as binary fractions rather than decimal fractions.

The well-ordering theorem in its general form was proved in 1904 by Hilbert's colleague Ernst Zermelo (1871–1953). However, his proof relied on the so-called axiom of choice. This axiom states that given a possibly infinite family of non-empty disjoint sets there is a set containing exactly one element from each of the sets in the family. This axiom, which is today accepted by the majority of mathematicians, may sound innocent, but it turned out to have a series of paradoxical consequences. The Banach–Tarski paradox is probably the most notorious: Given a sphere in space, it is possible to cut it into a finite number of disjoint pieces that can be assembled into two copies of the original sphere. The paradox was published in 1924 by the Polish mathematicians Stefan Banach (1892–1945) and Alfred Tarski (1901–83). Because of its paradoxical consequences, the axiom of choice was, and still is, rejected by some mathematicians, and many mathematicians will try to avoid using it if possible.

In 1908 Zermelo published a list of axioms for set theory, including the axiom of choice. This axiom system was modified by Abraham Fraenkel (1891–1965) in 1922. The so-called ZFC (Zermelo–Fraenkel (choice)) axiom system is still the preferred axiom system for set theory. Therefore, in a sense, Zermelo solved the second part of Hilbert's first problem by showing that the well-ordering theorem is a consequence of these axioms. In fact, it turns out that the well-ordering theorem is equivalent to the axiom of choice. Moreover, in 1938 Kurt Gödel (1906–78) showed that the negation of the axiom of choice does not follow from the remaining axioms,[8] and in 1963 Paul Cohen (1934–2007) proved that the axiom of choice itself does not follow from the rest of the axioms. In other words, the axiom of choice (or equivalently the well-ordering theorem) is independent of other axioms of set theory. Thus, the well-ordering theorem cannot be proved from the more universally accepted axioms of set theory. In that sense, the well-ordering theorem turned out to be undecidable, just as the parallel postulate had turned out to be unprovable from the other axioms of geometry.

The continuum hypothesis turned out to have a similar status. In 1940 Gödel proved that the continuum hypothesis can be added to the other axioms of set theory without creating any inconsistencies than those that may already be built into ZFC. Cohen then completed the result in 1963 by showing that one can also add the negation of the continuum hypothesis to the system. Thus, Cantor's and Hilbert's plan to deduce the continuum hypothesis turned out to be unfeasible, if we axiomatize set theory by ZFC.

[8] If they are consistent (see Chapter 17).

When Hilbert formulated his first problem in 1900, set theory was not formulated precisely enough to allow an answer. However, the search for an answer led to an axiomatization of the theory within which it could be demonstrated that the continuum hypothesis could not be proved or disproved and the well-ordering theorem could only be proved if one accepts the axiom of choice. Hilbert himself was extremely important in the development of the axiomatization of mathematics. In fact, his second Paris problem concerned the axioms of arithmetic. We shall now turn to that question, the "solution" of which was an impossibility result that shattered Hilbert's optimistic axiom of the solvability of every problem and his visions of safeguarding mathematics through an axiomatization of its disciplines.

17
Hilbert and Gödel on Axiomatization and Incompleteness

17.1 The axiomatization of mathematics

As we saw in Chapter 15, the investigation of the status of the parallel postulate and the resulting invention of non-Euclidean geometry led to a reinterpretation of the nature of axioms. Traditionally axioms had been considered as elementary truths about abstract or ideal reality. Toward the end of the nineteenth century, mathematicians gradually began to consider them as in principle arbitrary starting points for our deductions. At the same time, it was realized that Euclid and his followers had implicitly taken many things for granted that had not been explicitly formulated in the axioms. For example already in the proof of the very first proposition of the *Elements*, Euclid had assumed that two particular circles had a point of intersection. This seemed plausible from the figure, but was, in fact, not guaranteed by any axiom.

For that reason, several Italian and German mathematicians tried to formulate a complete system of axioms from which one could deduce all the theorems of geometry without recourse to intuition or figures. The most famous attempt is due to Hilbert. The first edition of his *Grundlagen der Geometrie* (Foundations of Geometry) was published in 1899. Subsequently, it was published in seven extended and considerably altered editions during Hilbert's lifetime.[1] Instead of Euclid's five postulates and five common notions, Hilbert needed about 20 axioms in order to achieve his goal of a complete system.

His presentation differed in another fundamental way from Euclid's presentation. Euclid had started his *Elements* with a list of definitions explaining the objects of geometry. For example, he defined a line as "a breadthless length" and a straight line as "a line which lies evenly with the points on itself." Already Aristotle had pointed out that it is in principle impossible to define everything. In order to define the meaning of a word one must use other words.

[1] The 10th edition was translated into English.

A History of Mathematical Impossibility. Jesper Lützen, Oxford University Press.
 DOI: 10.1093/oso/9780192867391.003.0017

In order to define these other words one must use yet other words, etc., ad infinitum. Indeed, Euclid's definitions raises the question: what do "breadthless" and "evenly" mean? Hilbert chose to shortcut the infinite regress inherent in such definitions by operating with undefined or implicitly defined objects. His *Foundations of Geometry* began thus:

> Let us consider[2] three distinct systems of things. The things composing the first system, we will call points and designate them by the letters A, B, C, ... ; those of the second, we will call straight lines and designate them by the letters a, b, c, ... ; and those of the third system, we will call planes and designate them by the Greek letters α, β, γ, ... We think of these points, straight lines, and planes as having certain mutual relations, which we indicate by means of such words as "are situated," "between," "parallel," "congruent," "continuous," etc. The complete and exact description of these relations follows as a consequence of the axioms of geometry.
>
> (Hilbert 1899/1992, 3)

In other words, Hilbert never explained what a point or a line *is* and he never explained what the words between, etc., mean. "Points," "lines," "between," ... just mean things or notions that satisfy the axioms. In that sense the meaning of the words are given implicitly by the axioms. A pocket watch could be considered a point, if it together with many other things called points would satisfy the axioms. From plane projective geometry it was known that one could interpret the words point and line in different ways. Indeed, a principle of duality states that all theorems in this part of geometry remain true when one interchanges the words point and line. Thus, a dot could play the role of a line when the figure drawn with a ruler plays the role of a point. Hilbert even suggested that one could replace points, lines, and planes with "chairs, tables and beer mugs."

Thus, Hilbert's geometry dealt with undefined objects according to rules that were in principle chosen arbitrarily. The mathematical logician Bertrand Russell (1872–1970) described this new conception of mathematics thus:

> Pure mathematics consists entirely of assertions to the effect that, if such and such a proposition is true of anything, then such and such another proposition is true of that thing. It is essential not to discuss whether the first

[2] This is the usual translation. However, it misses the point. It suggests that there is something out there that we consider. Hilbert wrote: "Wir denken ...," i.e., "we think three distinct ... ."

> proposition is really true, and not to mention what the anything is, of which it is supposed to be true. [...] Thus mathematics may be defined as the subject in which we never know what we are talking about, nor whether what we are saying is true.
>
> (Russell 1901, 75)

This modern conception of mathematics raised a new problem, namely the problem of consistency. An axiomatic system is said to be consistent when its axioms do not contradict each other. That means more precisely that it is not possible from the axioms to deduce both a statement p and its negation $\neg p$. It is clear that we do not like systems with contradictions and from a formal point of view they are uninteresting. Indeed, assume that a theory contains a contradiction $p \wedge \neg p$. Then any other statement q of the system is provable by contradiction because its negation $\neg q$ implies the contradiction $p \wedge \neg p$.

As long as axiom systems are supposed to describe something real, its consistency can be considered to be a consequence of the consistency of the real world. However, when axiom systems are in principle chosen at will and they do not describe anything outside the system, it is not at all obvious that the system is consistent. In Chapter 15 we saw how the invention of non-Euclidean geometry for the first time raised the problem of consistency. Was the new geometry, where the parallel postulate had been replaced by its negation, really consistent? We also saw how Beltrami found a way to establish consistency. He built a model of non-Euclidean geometry inside Euclidean geometry. At least such a model guarantees relative consistency: If Euclidean geometry is consistent, non-Euclidean geometry is consistent as well.

Beltrami and his contemporaries were convinced of the consistency of Euclidean geometry, but Hilbert's axiomatization emphasized the question. Hilbert solved the problem by building a model of Euclidean geometry, in arithmetic of the real numbers. The model was more or less the analytic geometry known since Descartes and Fermat. Thus, Euclidean geometry is consistent if arithmetic is. But is arithmetic itself consistent?

17.2 Hilbert's second Paris problem

Hilbert raised the problem of consistency of arithmetic in the second Paris problem. His presentation of the problem gives a fine explanation of his modern axiomatic conception of mathematics, so it is worth quoting.

> When we are engaged in investigating the foundations of a science, we must set up a system of axioms which contains an exact and complete description of the relations subsisting between the elementary ideas of that science. The axioms so set up are at the same time the definitions of those elementary ideas; and no statement within the realm of the science whose foundation we are testing is held to be correct unless it can be derived from those axioms by means of a finite number of logical steps. Upon closer consideration the question arises: *Whether, in any way, certain statements of single axioms depend upon one another, and whether the axioms may not therefore contain certain parts in common, which must be isolated if one wishes to arrive at a system of axioms that shall be altogether independent of one another.*
>
> (Hilbert 1900 in Gray 2000, 250)

The wish for independence of the axioms is in a sense a requirement of simplicity. It is a modern version of Occam's razor. Hilbert had already in his *Grundlagen der Geometrie* given examples of such proofs of independence. Having addressed the problem of independence, Hilbert continued with the main problem of consistency:

> But above all I wish to designate the following as the most important among the numerous questions which can be asked with regard to the axioms: *To prove that they are not contradictory, that is, that a definite number of logical steps based upon them can never lead to contradictory results.*
>
> In geometry, the proof of the compatibility of the axioms can be effected by constructing a suitable field of numbers, such that analogous relations between the numbers of this field correspond to the geometrical axioms. Any contradiction in the deductions from the geometrical axioms must thereupon be recognizable in the arithmetic of this field of numbers. In this way the desired proof for the compatibility of the geometrical axioms is made to depend upon the theorem of the compatibility of the arithmetical axioms.
>
> On the other hand a direct method is needed for the proof of the compatibility of the arithmetical axioms. The axioms of arithmetic are essentially nothing else than the known rules of calculation, with the addition of the axiom of continuity. I recently collected them ... .[3] I am convinced that it must be possible to find a direct proof for the compatibility of the arithmetical

[3] Hilbert's axioms for arithmetic of the real numbers were published in 1900. They are still used today with minor modifications.

axioms, by means of a careful study and suitable modification of the known methods of reasoning in the theory of irrational numbers.

(Hilbert 1900 in Gray 2000, 250–1)

Having thus presented his problem of the consistency of the axioms of real numbers in arithmetic, Hilbert went on to emphasize the philosophical importance of the question. Since the time of Plato, philosophers and mathematicians had contemplated what it meant for a mathematical object or concept to exist. Hilbert wanted to shortcut the philosophical discourse. For him a mathematical theory exists if it is consistent, and an object within a theory exists if the assumption of such an object does not lead to a contradiction:

> To show the significance of the problem from another point of view, I add the following observation: If contradictory attributes be assigned to a concept, I say, that *mathematically the concept does not exist.* So, for example, a real number whose square is -1 does not exist mathematically. But if it can be proved that the attributes assigned to the concept can never lead to a contradiction by the application of a finite number of logical processes, I say that the mathematical existence of the concept (for example, of a number or a function which satisfies certain conditions) is thereby proved. In the case before us, where we are concerned with the axioms of real numbers in arithmetic, the proof of the compatibility of the axioms is at the same time the proof of the mathematical existence of the complete system of real numbers or of the continuum. Indeed, when the proof for the compatibility of the axioms shall be fully accomplished, the doubts which have been expressed occasionally as to the existence of the complete system of real numbers will become totally groundless. The totality of real numbers, i.e., the continuum according to the point of view just indicated, is not the totality of all possible series in decimal fractions, or of all possible laws according to which the elements of a fundamental sequence may proceed. It is rather a system of things whose mutual relations are governed by the axioms set up and for which all propositions, and only those, are true which can be derived from the axioms by a finite number of logical processes. In my opinion, the concept of the continuum is strictly logically tenable in this sense only. It seems to me, indeed, that this corresponds best also to what experience and intuition tell us. The concept of the continuum or even that of the system of all functions exists, then, in exactly the same sense as the system of integral, rational numbers, for example, or as Cantor's higher classes of numbers and cardinal numbers. For

I am convinced that the existence of the latter, just as that of the continuum, can be proved in the sense I have described; unlike the system of *all* cardinal numbers or of *all* Cantor s alephs, for which, as may be shown, a system of axioms, compatible in my sense, cannot be set up. Either of these systems is, therefore, according to my terminology, mathematically non-existent.

(Hilbert 1900 in Gray 2000, 251–2)

17.3 The foundational crisis

In Hilbert's lengthy quote in Section 17.2, he refers to the doubt that had been raised as to the existence of the complete system of real numbers. This was a reference to mathematicians like Leopold Kronecker (1823–91), who advocated for a finitistic version of mathematics in which Cantor's theory of infinite sets was banned. As we saw in Chapter 16, Hilbert was a strong defendant of Cantor's ideas and argued that a restriction to finitistic methods would leave too much traditional and fruitful mathematics at the wayside. Still, the finitists had reasons to doubt the Cantorian world. In fact, as Cantor himself knew very well, unrestricted and what has later been called "naïve" use of set theory can lead to paradoxes. One of the most famous paradoxes was published by Bertrand Russell in 1903: Consider the set *M* of all sets that do not have themselves as an element. Russell asked the question: Does *M* contain itself as an element? If it does, then *M* satisfies the property characterizing elements in *M*, and thus it is not an element of itself. On the other hand, if it does not contain itself as an element, then *M* does not satisfy the property, and thus it contains itself as an element. Either way, we reach a contradiction.

Such paradoxes made it clear that one had to be careful with the new set theory. In particular, one cannot consider the collection of all sets as a set, and one has to outlaw that a set can have itself as an element. This is what Hilbert emphasized at the end of the quote in Section 17.2. While he urged an axiomatization of Cantor's higher classes of numbers and cardinal numbers, he also made it clear that one cannot show the existence of the *set* of all those infinite sets. As we saw in Chapter 16, his colleague Zermelo took up the challenge of axiomatizing set theory, a task continued by Fraenkel and von Neumann.

The finitistic conception of mathematics was continued and radicalized by the Dutch mathematician Luitzen E. J. Brouwer (1881–1966). Beginning around 1910 he developed a so-called intuitionistic philosophy of mathematics in direct opposition to Hilbert's axiomatic and formalistic ideas and to the logisist approach developed by Bertrand Russell. The difference between

Hilbert and Russell was that the latter thought that one can build mathematics entirely on logic, whereas Hilbert believed that different mathematical theories demanded different axiom systems. Brouwer departed radically from both. According to Brouwer mathematics is a creation of our intuition. It is constructivist, in the sense that its concepts are constructed in a finite number of steps. Actual infinite sets have no place here, and the logic Brouwer used was a limited version of usually accepted logic. In particular, Brouwer rejected the use of the law of excluded middle. This law states that for every proposition, either this proposition or its negation is true. The rejection of the law of excluded middle meant that indirect proofs are not allowed in intuitionistic logic. Brouwer argued that such limitations would safeguard him from the inconsistencies built into formalist mathematics built on a theory of infinite sets.

Hilbert was strongly opposed to Brouwer's intuitionism, which he considered an attempt to cripple mathematics. Beginning in 1917 he embarked on a great mission to save his axiomatic approach to traditional mathematics. The main purpose was to show that the axiom systems of arithmetic and set theory were consistent so that these theories existed from his point of view. In the process, he realized that a rigorous treatment of such foundational questions required the development of a formal logic. Building on Russell's logicistic major work *Principia Mathematica* (1910–13), Hilbert and his collaborators Paul Bernays (1888–1977), Wilhelm Ackermann (1896–1962), and John von Neumann (1903–57) made great progress in the development of a formal presentation of mathematics. Influenced by Brouwer, Hilbert believed that it would be possible to deal with the question of consistency with finitary methods rather similar to Brouwer's intuitionistic logic. If he had succeeded, he would have saved his mathematical universe from Brouwer's criticism.[4] However, in 1930 the logician Kurt Gödel directed a decisive blow to Hilbert's foundational program.

17.4 Gödel's incompleteness theorems

The central idea of Hilbert's formalistic program for the foundation of mathematics was the axiomatization of every branch of mathematics: That means listing a finite system of axioms from which all theorems of the theory can be deduced. With the formalization of logic the deducibility requirement was

[4] On the foundational crisis see Ferreirós (2008). On Hilbert's program see Zach (2019).

sharpened into a requirement of completeness: An axiomatic system is said to be complete if for any meaningful statement in the theory (i.e., any statement formed according to the rules of the theory) one can prove either the statement or its negation. For example, the Austrian logician Kurt Gödel (Figure 17.1) proved in his Ph.D. thesis from 1929 that so-called first order logic is complete.

Figure 17.1 Kurt Gödel (1906–78)
The University Library of Princeton University

However, the following year Gödel discovered that it is impossible to find a complete system of axioms for arithmetic and he implied that a similar incompleteness theorem could also be proved for set theory.[5] Having referred to Russell's *Principia Mathematica* and Zermelo–Fraenkel's axioms for set theory as prime examples of formalized theories, Gödel continued:

> One might therefore conjecture that these axioms and rules of inference are sufficient to decide any mathematical question that can at all be formally expressed in these systems. It will be shown below that this is not the case, that on the contrary there are in the two systems mentioned relatively simple

[5] On Gödel's theorems see Raatikainen (2020) and Nagel and Newman (1958). Gödel's theorems are complex. A precise formulation and understanding require a much more formal treatment than the intuitive outline I have been able to provide in this section. For example, Gödel only proved the incompleteness of the axiomatic system "obtained when the logic of PM [Russel's and Whitehead's *Principia Mathematica*] is superposed upon the Peano axioms." However, he remarked that this restriction was dispensable in principle. Moreover, he assumed a strange kind of consistency called ω-consistency. This also turned out to be an unnecessary restriction. For a more careful but still popular introduction to Gödel's theorems see Franzén (2005) and Stillwell (2010).

> problems in the theory of integers that cannot be decided on the basis of the axioms. This situation is not in any way due to the special nature of the systems that have been set up but holds for a wide class of formal systems; among these in particular, are all systems that result from the two just mentioned through the addition of a finite number of axioms, provided no false proposition ... become provable owing to the added axioms.
>
> (Gödel 1931 in Heijenoort 1981, 597)

In the main part of his paper, Gödel constructed an example of such an undecidable statement. It was inspired by the famous liar's paradox. The statement "this statement is not true" is not true, because in that case it would be false. On the other hand, it cannot be false either, because then it would be true. This constitutes a paradox. Gödel's statement was obtained by replacing "true" by "provable." The formal formulation of the statement was inspired by Cantor's diagonal argument and built on an intricate enumeration of all statements and proofs in the theory. In the informal introduction to the paper, Gödel wrote about the statement: "We therefore have before us a proposition that says about itself that it is not provable" (Gödel 1931 in Heijenoort 1981, 598). If the theory is consistent, the mentioned proposition cannot be proved, for if it could it would be unprovable which is a contradiction. Its negation cannot be proved either, for if it could, the proposition would be provable which also leads to a contradiction. Remark, that this is not a paradox but merely shows that the proposition and its negation are both unprovable.

In the last section of his paper, Gödel outlined an argument for his second incompleteness theorem to the effect that the consistency of the axiom system is one of the propositions that cannot be proved or disproved within the system.

Gödel's incompleteness theorems were widely considered as challenges to Hilbert's (and Russell's) foundational programs. The first theorem showed that it is impossible to formalize an informal mathematical theory which is strong enough to contain arithmetic. Whichever system of axioms one choses, there will be meaningful propositions in the theory that cannot be decided on the basis of the axiom system itself. And worse, the consistency of the system cannot be established using the method of the theory itself (except when the system is inconsistent, in which case every proposition is provable, even consistency). Since the establishment of consistency had been Hilbert's defense against the criticism of the intuitionists, Gödel's results could be considered a deathblow to Hilbert's program.

Still, in his paper, Gödel did not go that far:

> I wish to note expressly that Theorem XI [that is the second incompleteness theorem] ... do[es] not contradict Hilbert's formalistic viewpoint. For this viewpoint presupposes only the existence of a consistency proof in which nothing but finitary means of proof is used, and it is conceivable that there exist finitary proofs that cannot be expressed in the formalism of P (or M or A) [Gödel's formal systems].
>
> (Gödel 1931 in Heijenoort 1981, 615)

The formalisms P, M, and A were variants of the formalism of Russel's *Principia Mathematica.* However, soon after the appearance of Gödel's paper, most mathematicians agreed that the finitary methods Hilbert had developed for his proof theory would in fact be expressible in Gödel's formalism. Thus, despite Gödel's conciliatory remark, his theorems were quickly interpreted as real problems for the formalistic (and logicistic) philosophy of mathematics. For example, in June of 1931 von Neumann wrote to the logician Carnap: "Gödel has shown the unrealizability of Hilbert's program."[6] Soon it was even realized that Gödel's theorem could be strengthened. In 1963 Gödel made the following addition to his 1931 paper when it was published in his collected works:

> In consequence of later advances, in particular of the fact that due to A. M. Turing's work a precise and unquestionably adequate definition of the general notion of formal system can now be given, a completely general version of Theorem VI and XI [the two incompleteness theorems] is now possible. That is, it can be proved rigorously that in *every* consistent formal system[7] that contains a certain amount of finitary number theory there exist undecidable arithmetic propositions and that moreover, the consistency of any such system cannot be proved in the system.
>
> (Gödel 1931 in Heijenoort 1981, 616)

The formulation and proof of Gödel's theorems required a subtle "attention to syntactic and semantic distinctions, [a] restriction to particular formal systems and [a] concern for relative rather than absolute undecidability" (Dawson 1984, 97). Formalists like von Neumann who understood and accepted the

[6] Quoted from Mancosu (1999), which gives a fine discussion of the immediate reception of Gödel's theorems, including quotes from original letters between the actors.

[7] Rather than just the systems P, M, or A.

subtleties accepted the theorems although they were detrimental to their program. At first Hilbert himself did not think that his foundational program was affected by Gödel's theorems, but in 1939 he and Bernays included the first complete proof of Gödel's second incompleteness theorem in the second edition of their *Grundlagen der Mathematik* (Hilbert and Bernays 1939). Other mathematicians or philosophers such as Ludwig Wittgenstein (1889–1951) and Bertrand Russell were skeptical about the importance of Gödel's theorems or even rejected them. Their failure to accept Gödel's theorems is usually explained by their lack of understanding of Gödel's proof. According to Gödel himself, Wittgenstein "advance[d] a completely trivial and uninteresting misinterpretation" (quoted in Dawson 1984, 95) of his theorems. In a letter to Menger he went even further:

> As far as my theorem about undecidable propositions is concerned it is indeed clear from the passages you cite that Wittgenstein did not understand it (or pretended not to understand it)... . the whole passage you cite seems nonsense to me.
>
> (quoted in Floyd 1995, 409)

Other historians of logic and philosophy have tried to read Wittgenstein in a more favorable way. In particular Floyd (1995) has tried to understand Wittgenstein's reaction to Gödel's theorems in the light of his long interest in other impossibility theorems, in particular the impossibility of trisecting an arbitrary angle by ruler and compass.

The uneven reception of Gödel's theorems and the ongoing discussion among historians about how to interprete the standpoints of the historical actors illustrate that by 1930 mathematical logic had reached a new level of sophistication. From then on, research in the foundation of mathematics and mathematical logic was mostly left to specialists. "Outside [the community of logicians] it is difficult to assess to what extent Gödel's results were known, much less accepted or understood" (Dawson 1984, 94). Most practicing mathematicians continued to subscribe to a philosophy close to Hilbert's formalism. They just became used to the fact that even innocent-looking problems in number theory may turn out to be undecidable and that we may someday reach a contradiction in some part of mathematics that will force us to reformulate our axioms or make them more precise.

But despite the initial confusion about their real meaning and importance, Gödel's theorems did not stop the progress of mathematical logic and the foundation of mathematics. As most other impossibility results they acted as an incentive to continued research. They were, in Dawson's words, "a reaffirmation of the creative power of human reason" (Dawson 1984, 97).

17.5 Hilbert's tenth Paris problem

In the quote in Section 17.4 from 1931, Gödel referred to a clarification by Alan Turing (1912–54) of the concept of decidability in a formal system. This clarification also led to a "solution" of Hilbert's tenth Paris problem in a way that Hilbert had not anticipated. His presentation of the tenth problem is the shortest of them all.[8] Here is the full wording:

> Given a Diophantine equation with any number of unknown quantities and with rational integral numerical coefficients: To devise a process according to which it can be determined in a finite number of operations whether the equation is solvable in rational integers.
>
> (Hilbert 1900 in Gray 2000, 261)

Diophantine equations are the subject of an important part of number theory. They are polynomial equations with integer coefficients in any number of unknowns. The problem is to find integer (or rational) solutions to the equation. For example, Fermat's equation

$$x^n + y^n = z^n$$

is a Diophantine equation of degree n in three unknowns. This type of equation derives its name from the Greek mathematician Diophantus of Alexandria (AD c.250) whose *Arithmetica* dealt with such equations. According to Hilbert's "axiom" of the solvability of every mathematical problem, it should be possible to determine the solutions to any Diophantine equation or to prove that there does not exist a solution. His tenth problem did not go quite that far. It only asked for a process to decide whether a given Diophantine equation has rational solutions. He did not require that the process would produce the solutions when it decided that they exist.

Hilbert's tenth problem is in a sense a generalization of Fermat's last theorem that deals with one particular family of Diophantine equations in three unknowns. In addition to the special results obtained concerning Fermat's last theorem (see Chapter 18), some general results were known at Hilbert's

[8] On Hilbert's tenth Paris problem see Gray (2000, 226–32) and Alexandrov (1971, 177–95).

time. For example, in 1768 Lagrange had shown how to solve Diophantine equations of degree two in two unknowns. However, even in degree three no general solution method was known. Asking for a procedure for all degrees and any number of unknowns was a great step.

It is not quite clear what kind of "process" (*Verfahren*) Hilbert had in mind. He probably thought of a uniform procedure that would work for any Diophantine equation. But what kind of procedure? Today we would think of a so-called algorithm. But in 1900 a precise concept of an algorithm or a decision procedure was not yet formulated. That happened in the 1930s where first Gödel (1933) and then independently of each other the American mathematician Alonzo Church (1903–95) and the British mathematician Alan Turing (both 1936) came up with different descriptions of computability. According to Turing it can be decided whether a problem can be computed by a theoretical computer called a Turing machine. Church and Turing proved that the three notions of calculability or decidability were equivalent. It soon became generally accepted that this concept of calculability or decidability was the correct formal definition. This is the so-called Church–Turing thesis.

Having developed this formalization of the concept of a decision procedure or algorithm, and with Gödel's theorem in mind, mathematicians began to wonder whether an algorithm like the one Hilbert asked for in his tenth problem might not exist at all, and if so, whether it would it be possible to prove the impossibility.

These questions were investigated by the American mathematician Julia Robinson (1919–85). She began her research in the early 1950s, and in 1961 she and her two compatriots Martin Davis (b. 1928) and Hillary Putnam (1926–2016) succeeded in showing that there is no general algorithm to decide whether a so-called exponential equation is solvable by integers. Diophantine equations turned out to be more difficult. Yet, after almost a decade of work, and based on a paper of Julia Roberts, the Soviet mathematician Yuri Matiyasevich (b. 1947) in 1970 succeeded in proving a similar result for such equations. Thus, contrary to what Hilbert had believed, there is no algorithm that can decide whether a given Diophantine equation has integer solutions. It is debatable whether this negative answer to Hilbert's tenth problem shows that his optimistic rejection of ignorabimus in mathematics was mistaken. On the one hand, the result shows that we cannot know a uniform way to solve Diophantine equations; we cannot even find a method that can decide whether there are integer solutions. On the other hand, the negative answer to the tenth problem highlighted the importance of impossibility results. If we follow Hilbert and accept an impossibility proof as a solution, the problem had "finally found a fully satisfactory and rigorous solution, although in another sense than that originally intended" (to quote Hilbert's remark on earlier impossibility theorems) (Hilbert 1900 in Gray 2000, 247).

17.6 Conclusion

A similar conclusion can be formulated concerning Gödel's incompleteness theorems. Hilbert had emphasized the importance of impossibility theorems because he believed that if we accept a proof of impossibility as a "solution" of a problem, all mathematical problems would be solvable. With this optimistic belief in the mathematician's abilities, he wanted to refute Du Bois–Reymonds pessimistic "ignoramus et ignorabimus." However, according to the usual interpretation, Gödel's impossibility results showed that Hilbert was wrong. If we want a rigorous and entirely certain presentation of mathematics we need, according to Hilbert and his followers, to axiomatize and formalize it. But it turns out that in any sufficiently strong formalized system there are problems that we cannot solve or prove unsolvable, and statements that we cannot prove or disprove. Moreover, Gödel's second incompleteness theorem showed that the goal of axiomatization and formalization, namely mathematical certainty, was unattainable because it is impossible to prove the consistency of such formalized systems finitarily within the system itself. These are the pessimistic conclusions supporting the ignorabimus thesis.

However, Hilbert continued to reject ignorabimus in the domain of mathematics. On September 8, 1930, he gave a talk to the Society of German Scientists and Physicians at its congress in his native town of Königsberg. He concluded with these famous words:

> We must not believe those, who today, with philosophical bearing and deliberative tone, prophesy the fall of culture and accept the ignorabimus. For us there is no ignorabimus, and in my opinion none whatever in natural science. In opposition to the foolish ignorabimus our slogan shall be: We must know—we will know.
>
> (Wir müssen wissen—wir werden wissen).[9]

It is hard not to interpret this repetition of his optimistic view of mathematics, as a direct answer to Gödel, who had given his first brief presentation of his

[9] A four-minute version of the end of Hilbert's talk was recorded shortly afterward. The recording and an English translation are available on the Web (Hilbert 1930).

first incompleteness theorem one day earlier during a discussion at the Second Conference on the Epistemology of the Exact Sciences, held in the same town on September 5–7. However, it is unlikely that Hilbert was present during the discussion.

I shall also twist the conclusion of this chapter so as to end on a positive note, more in line with Hilbert's optimism. Indeed, we can accept Gödel's impossibility results as *solutions* to Hilbert's problems concerning axiomatized and formalized systems. If we do that, these impossibility results demonstrate the power of mathematics, in particular its ability to investigate the limits of its own activity.

18

Fermat's Last Theorem

18.1 Fermat's contribution

Fermat's last theorem is one of the most often-cited impossibility theorems. It states that if n is a natural number greater than 2, it is impossible to find natural numbers a, b, and c satisfying Fermat's equation

$$a^n + b^n = c^n. \tag{18.1}$$

Pierre de Fermat (Figure 18.1) formulated this theorem in the margin of his own copy of Bachet's 1621 edition of Diophantus' *Arithmetica* (about AD 250). In problem II.8, Diophantus showed how to find Pythagorean triples, i.e., natural numbers[1] a, b, c that can be lengths of sides of a right-angled triangle. According to Pythagoras, they must satisfy the equation

$$a^2 + b^2 = c^2. \tag{18.2}$$

Already around 2000 BC, the Babylonians knew how to find Pythagorean triples. It is debated how they found them, but several historians of mathematics read the preserved columns on the relevant clay tablet as evidence that they used a method found in Euclid's *Elements* (theorem X.29). In a slightly modernized form this theorem states:

Let $m > n$ be two mutually prime natural numbers of different parity.[2] Then

$$a = m^2 - n^2, \; b = 2mn, \; c = m^2 + n^2 \tag{18.3}$$

is a Pythagorean triple whose sides are mutually prime. Such a triple is called a primitive Pythagorean triple.

[1] Diophantus allowed the sides a, b, c to be rational numbers, but that does not basically change the problem.

[2] Meaning that one is even and the other is odd.

A History of Mathematical Impossibility. Jesper Lützen, Oxford University Press.
 DOI: 10.1093/oso/9780192867391.003.0018

Figure 18.1 Pierre de Fermat (1601?–65)
Reproduced by Jim Høyer DMS

The proof is easy. Just insert the values (18.3) for a, b, and c and calculate.

Conversely it holds true, but Euclid did not prove it, that any primitive Pythagorean triple a, b, c can be generated as in (18.3) from two mutually prime natural numbers $m > n$ of different parity. Here is a proof:

If a, b, c satisfy (18.2) and are mutually prime, any two of them must be mutually prime. Thus either a or b is odd. Assume without loss of generality that a is odd. Then b cannot be odd because the sum of two odd squares cannot be a square (the sum is divisible by 2 but not by 4). Thus, b must be even (say $=2b'$) and c must be odd. We can now write the Pythagorean equation (18.2) on the form

$$4(b')^2 = (2b')^2 = c^2 - a^2 = (c+a)(c-a).$$

The last two factors are both even, so we can write in natural numbers

$$(b')^2 = \frac{(c+a)}{2} \cdot \frac{(c-a)}{2}.$$

The two factors on the right-hand side must be mutually prime, since a common factor would be a factor in their sum and their difference, i.e., in c and b, but they were assumed to be mutually prime. When, as in this situation, a product of two mutually prime factors is equal to a square, both factors must be squares.[3] Thus, we can find m, n so that

$$m^2 = \frac{(c+a)}{2},\ n^2 = \frac{(c-a)}{2}.$$

It is now easy to see that (18.3) holds true and that m and n are mutually prime. They are of different parity because otherwise the sum and difference of m^2 and n^2 (i.e., c and a) would be even, but they are odd.

Diophantus used a similar method to find his Pythagorean triples. Fermat's famous marginal note stated the impossibility of generalizing this to higher powers than 2:

> In contrast it is impossible to divide a cube into two cubes, or a fourth power into two fourth powers, or in general any power beyond the square in infinitum into two powers of the same degree; of this I have discovered a truly marvelous demonstration which this margin is too narrow to contain.
>
> (Diophantus/Fermat 1670, 61)

Fermat usually communicated his mathematical results to his colleagues in the form of letters. However, he never included this "last theorem" in its general form in any of them. It became known after his death when his son Samuel in 1670 published a new edition of Bachet's edition of Diophantus, including all of Fermat's marginal notes and comments. There is some disagreement as to the dating of the note. Most historians of mathematics assume that it is a rather early note from around 1638, when Fermat in his letters began to mention the impossibility of solving Fermat's equation (18.1) for n equal to 3 and 4. However, Goldstein (1995, 120) has pointed out that it could also be a late note from 1659 when Fermat returned to his number theoretical research. At any rate, the theorem is not called Fermat's last theorem because it was the last theorem he formulated, but because it was left without a proof long after his other theorems had been proved or disproved.

Indeed, most historians and mathematicians believe that Fermat did not have a correct proof of the theorem. In fact, it is most likely that he himself discovered a mistake in the argument he at first believed he had found. That would explain why he never mentioned the general theorem in his letters.

The special case $n = 4$ is related to another number-theoretical impossibility theorem that Fermat proved. In fact, it is the only number-theoretical

[3] Just write the two factors as a product of primes.

theorem for which Fermat left a proof. The theorem states that there exists no right-angled triangle with integer sides, whose area is a square number. In other words if a, b, and c are natural numbers and (18.2) holds, then the area $\frac{1}{2}ab$ cannot be a square number. Fermat's proof is an example of a method he called the method of infinite descent. Assuming that a, b, and c are the sides of a Pythagorean triangle with square area, he could find a smaller triangle with the same property and so on ad infinitum. But that is impossible since any decreasing sequence of natural numbers must end after a finite number of steps.

Fermat's proof was penned down as a marginal note to the last proposition of his copy of Diophantus. Since it is condensed and formulated in a now unfamiliar geometric language, I shall give a slightly modernized version of it (see, e.g., Weil 1984, 77).[4] It is based on the above theorem about the nature of Pythagorean triples.

Given a Pythagorean triple a, b, and c whose area $\frac{1}{2}ab$ is a square, if the sides a, b, and c have a common factor greater than 1, we can get a smaller Pythagorean triple with the same property if we divide all sides with the common factor. On the other hand, if the sides a, b, and c are mutually prime, we know from the above theorem about Pythagorean triples that there exist two mutually prime natural numbers m, n of different parity such that the sides a, b, and c are given by (18.3). The area is $mn(m+n)(m-n)$, where each factor is mutually prime to the other three. If the area is a square, each of the mutually prime factors must be squares, say,

$$m = x^2,\ n = y^2,\ m + n = u^2,\ m - n = v^2. \tag{18.4}$$

Here u and v must both be odd and mutually prime. Setting $z = uv$ we have

$$x^4 - y^4 = m^2 - n^2 = (m+n)(m-n) = u^2v^2 = (uv)^2 = z^2. \tag{18.5}$$

Moreover, from (18.4) we have

$$2y^2 = 2n = (m+n) - (m-n) = u^2 - v^2 = (u+v)(u-v).$$

Since both u and v are odd, $(u+v)$ and $(u-v)$ must both be even, but they cannot have other non-trivial factors in common than 2 because such a factor

[4] A faithful translation and explanation of Fermat's proof can be found in Goldstein (1995, 61–5).

would be a common factor in u and v. Thus, one of the numbers $(u+v)$ or $(u-v)$ is of the form $2r^2$ where r is odd and the other is of the form $4s^2$. Hence

$$u = \frac{(u+v)+(u-v)}{2} = r^2 + 2s^2$$

and

$$v = \frac{(u+v)-(u-v)}{2} = \pm\left(r^2 - 2s^2\right).$$

From these formulas and (18.4) we see that

$$x^2 = m = \frac{(m+n)+(m-n)}{2} = \frac{u^2+v^2}{2} = \frac{\left(r^2+2s^2\right)^2+\left(r^2-2s^2\right)^2}{2} = r^4 + 4s^4.$$

Thus, x, r^2, and $2s^2$ are sides of a new Pythagorean triangle whose area r^2s^2 is a square. And the hypotenuse x is smaller than the hypotenuse a of the original triangle. This completes the proof by infinite descent.

In letters to his fellow mathematicians Fermat claimed that he had proved the impossibility of Fermat's equation (18.1) when $n = 3$ and $n = 4$ by the method of infinite descent. It is generally believed that he used a variation of the above argument to prove the impossibility of solving the equation

$$x^4 - y^4 = z^2, \tag{18.6}$$

which he had encountered in Eq. (18.5). This, in turn, shows that Fermat's equation (18.1) for $n = 4$ is impossible since a solution a, b, c to this equation would yield a solution to (18.6) by setting $x = c$, $y = a$ and $z = b^2$. However, Fermat left no clue as to the nature of his proof in the case $n = 3$. It lasted another century before Euler in 1753 succeeded in proving Fermat's last theorem for $n = 3$.[5] In this connection he remarked that his proof was so different from the proof in the case $n = 4$ that he saw no possibility of generalizing them to higher values of n.

Why did it take a century before new light was cast on Fermat's last theorem? To be sure, the problem is difficult, but Euler's proof did not require other methods than those available to Fermat. The reason must instead be sought in the limited interest in number theory in the seventeenth and eighteen8th centuries. Only a few of Fermat's correspondents shared his interest in questions

[5] There was a small hole in the argument.

about numbers, and those of them who did, such as Bernard Frénicle de Bessy (*c.*1604–74), were mostly interested in positive results. Negative results or impossibility statements like Fermat's last theorem were neglected or frowned upon.[6] For example, Wallis wrote about Fermat's negative propositions: "I do not see why he presents them as things of amazing difficulty. It is easy to imagine innumerably many assertions of the same kind."[7] In his letters to his colleagues, Fermat mentioned three negative statements about natural numbers in addition to his last theorem for $n = 3, 4$ (Weil 1984, 113). However, as pointed out by Goldstein, in his letters to his fellow mathematicians he always took care to mention the negative statements after the positive ones (Goldstein 1995, 134). He seems to have been aware of the low value commonly attributed to impossibility results.

18.2 Nineteenth-century contributions

During the end of the eighteenth and the beginning of the nineteenth century Lagrange, Legendre, and Gauss made great advances in number theory that the latter proclaimed to be the *queen of mathematics*. However, of these three only Legendre showed any interest in Fermat's last theorem. Gauss, in particular, echoed Wallis's negative opinion:

> I confess that Fermat's Last Theorem, as an isolated proposition, has very little interest for me, because I could easily lay down a multitude of such propositions, which one could neither prove nor dispose of.
>
> (Gauss 1816)

The first advances beyond Fermat's and Euler's proofs in the cases $n = 4$ and $n = 3$, respectively, were made by a woman, Sophie Germain (1776–1831). Even in revolutionary France, women could not attend lectures at the universities or other institutions of higher learning. Yet Germain succeeded in mastering mathematics by studying many of the mathematical textbooks of the time on her own. In particular, she studied Legendre's *Théorie des Nombres* when it appeared in 1798 and Gauss' masterpiece *Disquisitiones Arithmeticae* (1801). She became acquainted with Lagrange and Legendre and in 1808 began a correspondence with Gauss. All of these dignitaries were highly

[6] See Weil (1984, 114) and Goldstein (1995, 134, 146).
[7] Translated from Goldstein (1995, 134).

impressed by her mathematical originality. Already in her first letter to Gauss, she presented a proof of Fermat's last theorem for certain values of n. However, her proof turned out to be incorrect. She intensified her work on Fermat's last theorem in 1816 when the Academy of Sciences in Paris chose it for its prize problem for the year 1818. She never submitted a memoire to compete for the prize and indeed never published her extensive research on the subject. However, Legendre in a footnote in a paper of 1823 acknowledged her results and in particular a theorem now named after her.

Before we state Sophie Germain's theorem we should remark that if Fermat's last theorem is proved for a particular exponent n, it is a fortiori also established for exponents that are multiples of n. Indeed, a knth power is also an nth power. Thus, after Fermat had proved the theorem for $n = 4$, it would suffice to prove the theorem for all odd primes. Euler had disposed of $n = 3$, so the next cases in line were $n = 5, 7, 11, \ldots$.

Rather than trying her hand at one particular exponent, such as 5, Germain wanted to find general theorems that would allow her to prove Fermat's last theorem for all or at least for many exponents at once. In a letter to Gauss of 1819 she proposed such a general result:

Assume $a^n + b^n = c^n$ where n is an odd prime and that for some integer N, $2Nn+1$ is an auxiliary prime such that the set of non-zero nth power residues x^n $(x = 1, 2, 3, \ldots, n - 1)$ mod $(2Nn+1)$ does not contain any consecutive integers. Then the auxiliary prime will divide a, b, or c.

She then observed that if for a particular exponent n she could find an infinity of such auxiliary primes, then at least one of the numbers a, b, c would be divisible by an infinity of primes and that would be absurd. Unfortunately, Germain was unable to find such an infinity of auxiliary primes. However, she believed she could with similar methods prove that any solution to Fermat's equation for exponent 5 must have at least 39 digits.

The theorem that Legendre attributed to Germain says:

Sophie Germain's Theorem: For an odd prime exponent n, if there exists an auxiliary prime θ such that there are no two nonzero consecutive pth powers modulo θ, nor is p itself a pth power modulo θ, then in any solution to the Fermat equation $a^n + b^n = c^n$, one of a, b, or c must be divisible by n^2.

This shows that if one can find an auxiliary prime θ (just one!) of this kind, one has proved what is called Case 1 of Fermat's last theorem for that exponent. Case 1 says that if Fermat's equation (18.1) has a solution for an exponent n then a, b, or c is divisible by n.

Germain was able to find auxiliary primes of this kind for all values of the exponent n up to 100, and Legendre continued the list to 193. Thus, with Germain it was known that Case 1 of Fermat's last theorem holds true for odd prime exponents less than 197.

In her letters and manuscripts Germain went much further developing what Laubenbacher and Pengelley (2010) have called "a fully-fledged, highly developed, sophisticated plan of attack on Fermat's Last Theorem." And yet, she never succeeded in proving the impossibility of Fermat's equation for any new exponent.

In 1825 the German mathematician Peter Gustav Lejeune Dirichlet (1805–59) presented a paper to the French Academy in which he gave a proof of (half of) Fermat's last theorem for $n = 5$. For this exponent it is easy to see that the auxiliary prime $\theta = 2 \cdot 5 + 1 = 11$ satisfies the conditions of Germain's theorem. Thus, Case 1 is established and 5 must divide a, b, or c. It is easy to see that 2 must also divide a, b, or c. Dirichlet now proved that it is impossible that 5 and 2 divide the same of the three numbers. Later the same year Legendre showed that 2 and 5 could not divide two different of the three numbers either. His somewhat complicated proof was cleared up by Dirichlet one month later. Thus, Fermat's last theorem was established for exponent 5 in a common effort by Germain, Dirichlet, and Legendre.

Seven years later Dirichlet could prove Fermat's last theorem for $n = 14$. By publishing a proof for exponent 14, Dirichlet implicitly acknowledged that he could not manage the natural next step, exponent 7, which would have implied the weaker result proved by him. This natural next step $n = 7$ was taken by the French mathematician Gabriel Lamé (1795–1870) in 1839. Dirichlet's and Lamé's proofs are very technical and use methods special to those exponents. They did not point in the direction of a general method.

Thus, it came as a great surprise when on March 1st 1847 Lamé at a meeting of the French Academy announced that he had found a general proof of Fermat's last theorem. The main idea in Lamé's proof was to factor the left-hand side in Fermat's equation (18.1) over the complex numbers. If n is a prime and r is a imaginary nth root of unity (i.e., if $r^n = 1$), we can write

$$a^n + b^n = (a+b)(a+rb)(a+r^2b)\cdots(a+r^{n-1}b).$$

In the known proofs for n=3, 4, 5, 7, it was used that if a product of mutually prime numbers is an nth power, then each of the factors must be an nth power. Lamé wanted to use a similar theorem for complex factors of the above kind (so-called cyclotomic numbers). In this case, one can also introduce a concept similar to prime numbers. However, as Liouville pointed out after Lamé's talk, a generalization of the theorem to such complex factors would require that one could use the usual rules for prime factorizations, in particular that they are unique. It was not clear to Liouville that this uniqueness holds true for cyclotomic numbers and therefore he questioned the viability of Lamé's idea.

This incidence led to a feverish activity in the French Academy. Both Lamé and Cauchy continued over the next weeks to present notes (sometimes in secret envelopes) on Fermat's last theorem. However, at the meeting of May 24 Liouville read a letter from the German mathematician Ernst Kummer (1810–93) in which he pointed out that already three years earlier he had published (in a rather obscure journal) that for $n = 37$ unique factorization fails. After this blow to his great scheme, Lamé stopped his stream of notes and after some more months, Cauchy also gave up.

In his letter to the French Academy, Kummer added that one can save the theory of factorizations of cyclotomic numbers if one introduces what he called ideal complex numbers.[8] Using these new ideal numbers Kummer already in April of the same year proved that if two conditions were satisfied for the exponent n, Fermat's last theorem hold true for that value of the exponent. He then developed methods for testing whether a prime satisfies the two conditions (the so-called regular primes) and in this way succeeded in proving Fermat's last theorem for all primes below 100 except for the irregular primes 37, 59, and 67.[9] For his great advances with Fermat's last theorem, the French Academy in 1857 awarded Kummer its great prize that had been posed in 1853 for the proof of Fermat's last theorem.

18.3 The twentieth-century proof

While Kummer's results and his very difficult methods of proof gave Fermat's theorem its hitherto greatest leap forward, they also left a feeling among mathematicians that a complete proof would require enormous efforts. Most professional mathematicians considered it so difficult that they did not want to invest their energy in a research project that would probably result in failure. When Hilbert in 1900 formulated his famous list of important problems

[8] The theory of Kummer's ideal numbers is very complicated and its use in the solution of Fermat's last theorem is far from trivial. See Edwards (1977).

[9] Kummer even announced that there are an infinity of regular primes, but he later retracted the claim, which is still unproven.

that he believed would shape mathematics in the following century (see Chapter 16) he did not include Fermat's last theorem as a separate problem. Still he guessed correctly that it might find its proof in the following century.

The fame of the theorem got a public boost in 1908 when the industrialist Poul Wolfskehl left 100,000 Mark (more than one million dollars in modern-day money) for a prize that would go to the person who first proved Fermat's last theorem. It was up to the Royal Society of Science in Göttingen to decide whether a proof was complete. Thousands of "proofs" were sent to the society but they were mostly nonsensical or contained simple errors. During the twentieth century the theorem was established for more and more irregular prime exponents, i.e., primes that do not satisfy Kummer's two conditions. Since it required many long calculations, the endeavor was greatly facilitated by increasingly strong electronic computers that were built after the Second World War. In the 1980s it was established that the theorem held for exponents up to 25,000 and later this number was raised to several millions. But no matter how many exponents computers could dispose of, it was only a finite number, and an infinity of exponents still remained.

Wolfskehl must have been quite convinced of the validity of Fermat's last theorem. At any rate he did not mention what should happen if someone found a solution to Fermat's equation for some exponent greater than 2. Such a counterexample would have proved Fermat wrong but even computers could not come up with a counterexample.

After Gödel had proved his incompleteness theorem (1931) (see Chapter 17) there was even a third possibility: That the theorem had no counterexamples but could not be proven within the usual axiom system of arithmetic.

Figure 18.2 Andrew Wiles (b. 1953)
Photograhed by Jim Høyer, DMS

Toward the end of the twentieth century Fermat's last theorem was finally proved by the English mathematician Andrew Wiles (Figure 18.2). His long and complicated proof was based on the profound work of several other twentieth-century mathematicians concerning elliptic curves and modular forms. An elliptic curve is an algebraic curve with an equation of the form

$$y^2 = ax^3 + bx^2 + cx + d,$$

where a, b, c, d are rational numbers. The properties of rational points (x, y) on such curves had been studied in the eighteenth century, and by the late nineteenth century it was known that one could define a notion of addition that would make the rational points into a group. Modular forms, on the other hand, are certain complex functions mapping the positive complex half-plane into the complex plane. Modular forms also come with a group structure. In the mid-1950s two Japanese mathematicians, Yutaka Taniyama (1927–58) and Goro Shimura (1930–2019), became convinced that there was a deep connection between the two different subjects: Every elliptic curve comes in a precise way from a modular form. The conjecture was made more precise by the French mathematician André Weil (1906–98) and was a cornerstone of a wide program of unification of different fields of mathematics devised by Robert Langlands (b. 1936). The connection of these ideas with Fermat's last theorem was conjectured by Gerhard Frey (b. 1944) in the early 1980s and clarified by Jean-Pierre Serre (b. 1926) and Kenneth Ribet (b. 1947) in 1986. They showed that if a solution to Fermat's equation could be found, one could construct an elliptic curve possessing such strange properties that it could not come from a modular form. Thus, such a solution would imply that the Taniyama–Shimura conjecture would be false. Conversely, if one could prove the Taniyama–Shimura conjecture one would have proved Fermat's last theorem.

In 1986 the general feeling was that if the Taniyama–Shimura conjecture was true it would be exceedingly difficult to prove it. But the young Andrew Wiles, who had been fascinated by Fermat's last theorem since his childhood, decided to use all his efforts to proving the conjecture. He was well equipped for the task after having studied elliptic curves for his Ph.D. at Cambridge University. After defending his thesis, he moved to Princeton where he worked as a professor and where he heard about the proof of Frey's conjecture in 1986. He immediately decided that he would fulfill his childhood dream by proving the Taniyama–Shimura conjecture and thus Fremat's last theorem. He worked on the proof in complete secrecy and isolation for seven years before he had completed his proof. On June 21–3, 1993 Wiles gave a series of three talks entitled "Modular Forms, Elliptic Curves and Galois Representations" at a number theory meeting in Cambridge. Already after the first talk, it was rumored that he would prove the Taniyama-Shimura conjecture. Thus, the lecture hall was packed when he ended his proof of a special case of the conjecture that included the Frey curve and concluded by writing Fermat's last theorem on the blackboard saying: "I think I will stop here."

He sent his proof to the journal *Annals of Mathematics*, but during the fall, the reviewers discovered a flaw in the proof. Wiles spend the next year trying to mend it and finally in September 1994 he and his former Ph.D. student Richard Taylor (b. 1962) found a way out. The amended proof was published the following year in two long papers in *Annals*, finalizing the 350-year long quest. By its very nature, Wiles' twentieth-century proof cannot be anything like the argument Fermat at one time thought he had found.

The proof of Fermat's last theorem won Wiles the Wolfskehl prize, and he would certainly also have won the most prestigious mathematics award, the Fields Medal, had it not been because this prize had an age limit of 40 years that Wiles had just past when he finalized the proof. In addition to solving one of the most famous riddles of mathematics, Wiles' proof also developed new methods that have been important in other areas of algebraic geometry and number theory. This is representative for the whole history of Fermat's last theorem: The theorem itself may not have been so important, but as a result of its great difficulty, it has forced mathematicians to develop many deep methods that have been important for the study of other subjects as well. This was already emphasized by Hilbert in his Paris lecture on mathematical problems in 1900 where he said about Fermat's last theorem:

> The attempt to prove this impossibility offers a striking example of the inspiring effect which such a very special and apparently unimportant problem may have upon science. For Kummer, spurred on by Fermat's problem,[10] was led to the introduction of ideal numbers and to the discovery of the law of the unique decomposition of the numbers of a cyclotomic field into ideal

[10] Edwards (1977, 79) argues convincingly, that it was in fact Jacobi's work on higher reciprocity laws that originally led Kummer to his ideal numbers.

prime factors—a law which today, in its generalization to any algebraic field by Dedekind and Kronecker, stands at the center of the modern theory of numbers and the significance of which extends far beyond the boundaries of number theory and into the realm of algebra and the theory of functions.

(Hilbert 1900, 242 in Gray 2000)

19
Impossibility in Physics

19.1 The impossibility of perpetual motion machines

In Chapter 16 we saw how Hilbert emphasized the great importance of impossibility theorems. The recent progress in this area had convinced him that if we accept a proof of impossibility as a solution of a problem, then all mathematical problems would be solvable. In mathematics there would be no ignorabimus. In a short passage of his 1900 Paris lecture he extended this argument to physics and other realms of human investigation:

> Is this axiom of the solvability of every problem a peculiarity characteristic only of mathematical thought, or is it possibly a general law inherent in the nature of the mind, a belief that all questions which it asks must be answerable by it? For in other sciences also one meets old problems which have been settled in a manner most satisfactory and most useful to science by the proof of their impossibility. I cite the problem of perpetual motion. After seeking unsuccessfully for the construction of a perpetual motion machine, scientists investigated the relations which must subsist between the forces of nature if such a machine is to be impossible, and this inverted question led to the discovery of the law of the conservation of energy, which, again, explained the impossibility of perpetual motion in the sense originally intended.
>
> (Hilbert 1900 in Gray 2000, 248)

Thus, for Hilbert, impossibility results were also important outside of mathematics and seemed to suggest the general validity of the axiom of the solvability of every problem. In physics and other natural sciences, an impossibility like other results can arise in two ways. It can have an empirical origin or it can be a result of the theoretical, usually mathematical description or model of the science in question. However, as indicated by the highly rationalized historical example given by Hilbert, impossibilities in physics often arise from a combination of empirical and theoretical considerations.

Indeed, in a mathematical theory (or model) of physics or any other science, impossibilities can be established by mathematical methods, just as in

A History of Mathematical Impossibility. Jesper Lützen, Oxford University Press.
 DOI: 10.1093/oso/9780192867391.003.0019

a theory of pure mathematics. All impossibility results of a mathematical theory, including fundamental results like Gödel's theorems, also hold in sciences that are described using this theory. Moreover, in applied mathematics the mathematical formalism is supposed to describe something real. In formalized pure mathematics, the axioms of the theory are chosen at will, subject only to the requirement of consistency, and the fundamental objects are only defined implicitly by the axioms. When a mathematical theory is applied to science, the objects must correspond to observable things of reality and the axioms or principles must be empirically true of reality. Thus, impossibility of a mathematized area of science can also arise from empirical measurements. But the distinction between empirical impossibility and mathematical impossibility is blurred. Indeed, if we want to give a mathematical description of a scientific discipline, we must choose some of its principles as axioms from which the other results can be deduced. We can then think of the axioms as the empirical elements of the theory and the other results as mathematical consequences. But there is no unique choice of axioms. Many different choices can result in the same theory. For example, if we think of geometry as a description of physical space, we can base it on the parallel postulate or the statement that the angle sum in a triangle is 180°. In that case one of the two is an empirically based axiom and the other is a mathematical consequence.

To make matters even murkier, Pierre Duhem (1861–1916) and Willard Van Orman Quine (1908–2000) have pointed out that we can never test an isolated theorem in a mathematized theory of reality for its empirical correctness (the Quine–Duhem thesis). A good illustration of this is due to Poincaré. Around 1900 when he gave the example, all measurements indicated that Euclidean geometry was a good mathematical model of physical space. However, imagine that at some point in time one measures a triangle in which the angle sum is less than 180°. Would that prove the impossibility of Euclidean geometry, in particular the parallel postulate in the description of physical space? Not at all! Indeed in order to measure angles in large triangles one would have to use light rays to form the sides of the triangle. Therefore, if the angle sum in such a triangle of light rays deviates from 180°, there are two possibilities: Either geometry is not Euclidean or light does not follow straight lines. If light does not follow straight lines, then either light is not composed of electromagnetic waves or the electrodynamic fundamental equations, Maxwell's equations, are not accurate. It is a matter of convenience where one makes alterations to the theory in order to make it agree with experience. The conclusion of this analysis is that one can only test a whole theory for its empirical ability to describe reality; one cannot test single components.

That also holds for impossibility results, as illustrated by the example discussed by Hilbert: The impossibility of a perpetual motion machine.

A perpetuum mobile or perpetual motion machine is a machine that can run forever without being driven by any external energy source. The dream of such a machine arose during the Middle Ages as a byproduct of the invention of many mechanical machines such as windmills. The ultimate perpetuum mobile was supposed to be able to perform work while it was running. Many different designs were constructed on paper and in reality, but none of them worked. As remarked by Simanek (2020), "even in the earliest history of science and engineering, many persons were able to see the futility and folly of attempts to achieve perpetual motion." For example, Leonardo da Vinci (1452–1519) wrote in one of his note books: "Oh, ye seekers after perpetual motion, how many vain chimeras have you pursued? Go and take your place with the alchemists" (McCurdy 1906, 64). The widespread conviction of the impossibility of a perpetual motion machine was probably not only based on the failing of all attempts but also a result of a sound common sense (or philosophically supported) conviction that one cannot get something from nothing. The impossibility rested on experience and common sense. The failing of many purported perpetual motion machines could be demonstrated using Archimedean methods of statics (the law of the lever) but a general theoretical proof of the impossibility of any kind of perpetual motion machine was only achieved during the nineteenth century. Yet, when the French Academy of Sciences in 1775 declared that they would no longer evaluate circle quadratures, angle trisections, and cube duplications, they also included perpetual motion machines as impossible constructions they would no longer deal with.

The basis for a theoretical proof of the impossibility of perpetual motion machines lies in the concepts of energy and entropy. The concept of energy has its origin in the seventeenth-century discussions between continental and British natural philosophers about the correct measure of motion. Where Newton preferred what we today call momentum, i.e., the product of mass and velocity (a directed quantity or vector in modern terminology), Leibniz opted for the living force (force vive) which is the product of mass and velocity squared. From Galileo's investigations of free fall, it was known that the force vive of a falling body was proportional to the height it has fallen. This observation was generalized toward the end of the eighteenth century when French mathematicians showed that the usual forces of nature, like gravity and electric and magnetic force, could be found as the gradient of a potential function. It was shown that when a mechanical system moved under the influence of such forces without friction, the gain in half of the force vive was equal to the increase in the potential function. During the first half of the nineteenth century this led to a shift in the sign of the potential function, leading to the concept named potential energy by William Thomson (Lord Kelvin) (1824–1907). Simultaneously the force vive (multiplied by ½) changed name to kinetic energy such that the law of conservation of the sum of the potential and the kinetic energies was formulated.

In the meantime, experiments with heat revealed that it could be considered as arising from the motion of microscopic constituents of matter. In the 1840s Julius Robert von Mayer (1814–78), Ludvig August Colding (1815–88), and most famously James Prescott Joule (1818–89) determined the mechanical equivalent of heat, showing thereby that heat was a kind of energy. Finally, in 1847 Hermann von Helmholtz published a book entitled *Über die Erhaltung der Kraft* in which he argued that in an isolated system the energy (which he still called *Kraft* or force) is always conserved. He even showed that this principle applied to living systems as well. As a law about thermodynamic systems, such as steam engines, this law is usually called the first law of thermodynamics.

The principle of energy conservation tells us that a perpetual motion machine is impossible. If it is to perform work, it will have to suck energy out of some other system. And since friction is unavoidable it cannot run for ever, even if it does not produce useful work. Energy conservation does not in itself exclude that an engine could suck thermal energy out of a reservoir by cooling it (a perpetuum mobile of the second kind). But that is impossible by the second law of thermodynamics. This law was formulated by Rudolf Clausius (1822–88) in the 1850s on the basis of Sadi Carnot's work from the 1820s. It concerns a quantity called entropy that was later given, by Maxwell and others, a statistical interpretation as a measure of disorder in a system. Clausius showed that in an isolated thermodynamic system the entropy always increases. In a machine, it will decrease. That means that a thermodynamic machine can only work if it has two reservoirs of different temperature. While the machine runs, it will transfer heat from the hotter to the colder reservoir. It will stop working when the two reservoirs have the same temperature. Thus, a perpetual motion machine of the second kind is also impossible.

This brief and highly simplified story tells us that the impossibility of a perpetual motion machine began as an empirical (or common sense philosophical) result but ended up as a consequence of two more fundamental principles, namely the conservation of energy and the increase of entropy. In fact, the conservation of energy was later shown to be a consequence of an even deeper principle of physics, Noether's theorem. This theorem was proved in 1915 and published in 1918 by Hilbert's colleague Emmy Noether

(Figure 19.1), often described as the most important woman in the history of mathematics. Noether's theorem states that to each symmetry of a physical system there corresponds a conservation law. For example, if a system has rotational symmetry, its angular momentum is conserved. The conservation of energy is a consequence of the symmetry of the system under time translation; i.e., it is a consequence of the fact that the laws of motion are the same for all times.

Figure 19.1 Emmy Noether (1882–1935)
Reproduced by Jim Høyer DMS

As we argued earlier, the changing nature of the impossibility of a perpetual motion machine is not just a result of a historical development. It is inherent in the nature of mathematical descriptions of reality. Indeed, even in modern texts on thermodynamics one sometimes encounters the impossibility of a perpetual motion machine as an empirically based axiom or principle from which energy conservation follows as a consequence.

19.2 Twentieth-century impossibilities in physics

Modern physics of the twentieth century witnessed many important impossibility statements in physics. Here we shall only present a few of them. Both quantum mechanics and relativity theory were born as a result of the impossibility of explaining new experimental results using the classical physics of the nineteenth century.

The atomic model suggested by Niels Bohr (1885–1962) in 1913 postulated that electrons circulate in stationary orbits around the nucleus of an atom. Such a stationary motion would be impossible according to classical electrodynamics, according to which an electron emits electromagnetic radiation and spiral in toward the nucleus. According to Bohr, electrodynamic radiation is only emitted when the electron jumps from one stationary orbit to another. So in a sense the old quantum mechanics was based on an assumption of the possibility of the impossible. With the advent of quantum mechanics proper in 1925, a more fundamental impossibility crept into physics, namely the impossibility of measuring the value of the observables of a physical system as accurately as we want. The assumption of accurate measurability was a cornerstone of the classical deterministic paradigm formulated around 1800. At that period of enlightenment and revolution, the great success of classical mechanics, in particular its accurate description of motion in the planetary system, led Laplace to conclude that it is, in principle, possible to predict the position of all particles in the universe for all times, if we know their positions and velocities at one particular moment. We "just" have to solve the equations of motion with given initial values. This is the paradigm of determinism. When we throw a die, the outcome is uniquely determined by the motion of our hand and the physical properties of the die and the table it lands on. However, Laplace acknowledged that humans will never be able to acquire precise knowledge of the initial conditions and physical properties; thus, we have to make due with probabilities when dealing with throws of a die. According to Laplace, probability theory is developed in order to get around human imperfection, not because nature is stochastic.

In quantum mechanics, however, it is in principle impossible to accurately measure the position and velocity of a particle simultaneously. The value of an observable like position and velocity is determined by a probability distribution, and when the distribution of position becomes narrower, so that we know the position more accurately, the distribution of the velocity becomes wider. The relation was expressed accurately by Werner Heisenberg (1901–76) in 1927 in his famous uncertainty relation

$$\Delta x \cdot \Delta p \geq \frac{h}{4\pi},$$

where h is Planck's constant and Δx and Δp are the standard deviations (the uncertainty) in determining the position x and the momentum p (velocity multiplied by the mass of the particle). That means that the initial state of a physical system can only be determined with some probability, and thus Laplace's deterministic paradigm was shattered. For example, when a radioactive atom decays, it is possible to calculate the probability with which it will decay in a particular time interval, but there is no way we can predict precisely when it will happen. The decay just happens spontaneously without something causing it to decay now rather than in 5 minutes. Quantum mechanics defies our usual ideas of causality.

The probabilistic interpretation of quantum mechanics was introduced by Max Born (1882–1970) in 1926. He suggested that the so-called wave function, the fundamental characteristic of a physical system in Schrödinger's version of quantum mechanics, determined the probability of finding the system at a particular place. According to the Copenhagen interpretation of quantum mechanics developed by Bohr and Heisenberg shortly afterward, there is no way to avoid the probabilistic nature of quantum mechanics. This interpretation was rejected by Einstein. Already in a letter to Born in 1926, Einstein famously wrote about "the old one" meaning God: "Quantum theory yields much, but it hardly brings us close to the Old One's secrets. I, in any case, am convinced He does not play dice with the universe." For the rest of their lives Bohr and Einstein continued to argue whether quantum physics is, in principle, of a statistical nature or whether the statistical elements are only apparent.

Already in the last half of the nineteenth century, James Clerk Maxwell (1831–79) and Ludwig Boltzmann (1844–1906) had developed a statistical approach to thermodynamics. Their idea was that the value of macroscopic variables such as pressure and temperature was the result of the motion of the molecules of matter. They believed that, in principle, the motion of the molecules making up the thermodynamic system, such as a steam engine or the atmosphere, could be determined in the way described by Laplace. However, in practice we are unable to measure the position and velocity of each molecule because they are too small, and there are too many of them. Moreover, in most cases we are not interested in in the motion of every single molecule, but only in the development of the macroscopic characteristic of the system such as temperature and pressure that reflect the statistical distribution of the molecules. The behavior of these macroscopic variables is described by thermodynamics.

With the rise of quantum mechanics, the question naturally arose: Is it possible that the apparent stochastic and non-deterministic character of the theory could be due to our lack of complete knowledge of the system? Could it be that just as temperature and pressure are, in fact, determined by a host of underlying variables (the positions and velocities of the molecules), the wave function does not give us a complete description of the quantum mechanical system, but requires other variables to determine its precise state? Could it be that if we take the hidden variables into account, the theory would become entirely causal and deterministic, so that we are back to the Laplacean paradigm: Probability is only needed because we do not know the system sufficiently well? Earlier physicists had introduced hidden variables or substances in order to account for the motion of mechanical systems. Descartes had believed that all interactions were due to impact between ordinary matter and an all-pervasive ether that was hidden from our senses. In the nineteenth century, the ether served as the substance in which light and electromagnetism could propagate. Lord Kelvin suggested that ordinary matter was simply vortices in the ether, and Heinrich Hertz (1857–94) suggested (1894) that forces could all be explained by rigid constraints and a system of hidden masses.

According to the Copenhagen interpretation of quantum mechanics, such a hidden-variable rescue of determinism is impossible. Bohr argued for it in rather philosophical terms[1] and in 1932, John von Neumann (Figure 19.2) provided a mathematical proof of the impossibility. Von Neumann's impossibility proof appeared in his famous book *Mathematische Grundlagen der Quantenmechanik* (Mathematical Foundation of Quantum Mechanics). According to Lacki (2000), von Neumann's book was a partial answer to Hilbert's call for an axiomatization of physics,[2] and his proof of impossibility was a proof of the completeness of his axiom system in the sense that the theory could not be extended to include hidden variables. Von Neumann founded quantum mechanics on a theory of operators in Hilbert space. It was mathematically rigorous as opposed to Dirac's earlier theory that had used a non-existing function, the delta function.

Von Neumann's impossibility proof has been the subject of much discussion.[3] Already in 1935 Grete Hermann (1901–84) challenged its conclusions but her papers had little influence. Indeed, until the 1960s von Neumann's proof was widely considered as the proof of the probabilistic Copenhagen interpretation of quantum mechanics. However, in 1952 David Bohm (1917–92) put forward an "interpretation of the quantum theory in

[1] For Bohr's philosophy of quantum physics see Murdoch (1987).

[2] The axiomatization of physical theories was presented as the sixth problem in Hilbert's 1900 Paris lecture.

[3] Accounts of von Neumann's proof and its reception can be found in Lacki (2000), Seevinck (2016), and Dieks (2017).

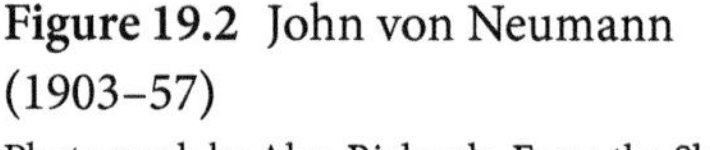

Figure 19.2 John von Neumann (1903–57)

Photograph by Alan Richards. From the Shelby White and Leon Levy Archives Center, Institute for Advanced Study (Princeton, N.J.)

terms of hidden variables," and in 1966 John Bell (1928–90) argued that "the realization that von Neumann's proof is of limited relevance has been gaining ground since the 1952 work of Bohm." Bell was a strong opponent of the Copenhagen school and succeeded in casting doubt on the correctness of von Neumann's impossibility proof. More recently, Lacki (2000) and in particular Dieks (2017) have argued that criticism of von Neumann's impossibility argument was misdirected. Bell and his followers argued that it is possible to introduce hidden variables in quantum theory, but von Neumann did not claim that his proof contradicted this possibility. What von Neumann proved to be impossible is the introduction of hidden variables in a quantum theory that is axiomatized the way von Neumann had done, in which observables are described by so-called Hermitian operators on Hilbert space. He explicitly mentioned the possibility that future experiments may lead to a different mathematical formalization of physics but continued:

> in the present state of our knowledge everything speaks against this: for the only presently available formal theory that orders and summarizes our experiences in a somewhat satisfactory way, namely quantum mechanics, is in strict logical conflict with [deterministic causality]. However it would be an exaggeration to claim that this is the absolute end of causality ... . One can never say of a theory that it has been proven by experience ... . But even considering all those words of caution we are allowed to say: at the moment there is no reason and no excuse to keep on talking about causality in nature—for no experience supports its existence, as macroscopic experience is unable to

> do so as a matter of principle, and the only known theory that is compatible with our experiences concerning elementary processes, quantum mechanics, contradicts it.[4]

Thus, von Neumann presented his proof as a purely mathematical proof concerning his axiomatic foundation of quantum mechanics. It is still hotly debated whether one can find a causal, deterministic theory using hidden variables that can describe microscopic physics.

Also relativity theory has its impossibility theorems, in particular the impossibility of motion with velocity greater than the velocity of light. However, it has also opened for a reversal of a famous and long-held impossibility, namely the impossibility of time travel. Indeed, in general relativity there is a theoretical possibility of traveling back in time. The physical possibility and the philosophical implications are still under debate.

Since this book is about impossibilities in mathematics, I shall not dwell any further on the impossibilities arising in connection with the application of mathematics to the physical sciences. Still let me conclude by quoting John Barrow, whose book (Barrow 1998) goes much deeper into the matter:

> Limits to what is impossible may turn out to define the universe more powerfully than the list of possibilities.
>
> (Barrow 1998, 249)

[4] Von Neumann (1932), as translated by Dieks (2017, 141), who offered a rehabilitation of von Neumann's proof.

20
Arrow's Impossibility Theorem

In the second half of the twentieth century, mathematical impossibility theorems began to play an important role in the social sciences. The first such theorem was Arrow's impossibility theorem that gave rise to the theory of social choice.[1] In this chapter, we shall see how the modern axiomatic approach to mathematics was decisive in Arrow's discovery of his theorem, just as it had been in von Neumann's proof of his impossibility theorem concerning quantum mechanics.

20.1 The theory of voting

Arrow's impossibility theorem was the result of developments in two almost disjoint areas: the theory of voting and welfare economics. In the theory of voting, the mathematical problem addressed by Arrow's theorem was discovered already in 1785 by the French mathematician Nicolas de Condorcet (1743–94) (Figure 7.2). He wrote the book entitled *Essay on the application of analysis to the probability of decisions made on a majority vote* (Condorcet 1785) in order to show that

> the truths of the moral and political sciences are susceptible to the same certainty as those that form the system of physical sciences and even the branches of the sciences which, as astronomy appears to approach mathematical certainty.
>
> (Condorcet 1785, i)

As an example, he chose the theory of voting. He gave a detailed account of the situation where a population had to decide between two alternatives. For our purpose, the interesting part of the book is the last part devoted to the situation where there are more than two alternatives to choose from. Condorcet immediately rejected the usual majority rule according to which the alternative getting most votes wins. This method is unsatisfactory, because it

[1] For a more detailed account of the history of Arrow's impossibility theorem see Lützen (2019).

A History of Mathematical Impossibility. Jesper Lützen, Oxford University Press.
 DOI: 10.1093/oso/9780192867391.003.0020

can easily happen that the chosen alternative is in fact the most unpopular, or would at least lose to other alternatives in a pairwise election between just those two. So, instead, Condorcet suggested that one should choose that alternative which would win over all the other alternatives in a pairwise election. Such a winner is now called a Condorcet winner.

However, Condorcet discovered that if there are more than two alternatives, it is not certain that there is a Condorcet winner. The simplest example (simpler than the examples put forward by Condorcet) is the situation where there are three voters who have to choose one among the three alternatives A, B, and C. Assume that voter number 1 orders the three alternatives A, B, C, meaning that he prefers A over B and B over C. Assume furthermore that voter 2 orders the alternatives B, C, A, and voter 3 orders them as C, A, B.

In that case, A will win in a pairwise election between A and B, because both voter 1 and 3 prefers A to B. Only voter 2 prefers B to A. Similarly in a pairwise election between B and C, B will win over C with the votes of voter 1 and 2. When A wins over B and B over C one would expect that A would win over C. However, in a pairwise election between A and C, C will in fact win over A with the votes of voter 2 and 3. This is the so-called Condorcet paradox.

In mathematical terms the problem in the above example is that although each of the three voters order the alternative in a transitive way, the Condorcet method ends up with an intransitive ordering. An ordering $\geq$ is called transitive if

$$x \geq y \textit{ and } y \geq z \textit{ implies } x \geq z.$$

This holds in the above example for each voter separately, but still the pairwise voting procedure leads to the intransitive (circular) situation $A \geq B \geq C \geq A$.

The question is whether one can find a voting procedure other than Condorcet's that does not have this or other undesirable properties. The problem of finding a good voting procedure had already been studied in the Middle Ages and continued after 1770 to be the subject of many rather unconnected investigations undertaken by engineers, statesmen, architects, lawyers, and mathematicians. Many of the proposed methods were applied in practice in all sorts of elections. The most systematical investigations prior to 1900 were undertaken by the mathematician Charles Dodgson (1832–98), better known under his pen name Lewis Caroll, the author of *Allice in Wonderland*.

The first person who devoted a majority of his mathematical energy to voting procedures was the Scottish mathematical economist Duncan Black (1908–91). In 1942 he discovered a situation where it would be easy to find a

Condorcet winner of an election. If the alternatives can be arranged in such an order that each voter has a preferred alternative and likes the other alternatives less the further they are from the preferred alternative, Black called it single peaked. Such situations are very common. For example, if the alternatives are numbers representing the price of a product to be marketed or the load of a particular tax, the voters will usually order their alternatives in a single peaked fashion. The same holds in a traditional parliamentary election when the parties can be ordered from left to right. In such situations the median peak will be a Condorcet winner. The median is the alternative preferred by one of the voters, such that there are equally many voters preferring alternatives below and above this one.

Black at first believed that even when the preferences are not single peaked, there would be a Condorcet winner. However, as he tried to find a general rule to pick out the winner he stumbled upon the Condorcet paradox. It was a very emotional experience for Black:

> Later in working out an arithmetical example, an intransitivity arose, and it seemed to me that this must be due to a mistake in the arithmetic. On finding that the arithmetic was correct and the intransitivity persisted, my stomach revolted in something akin to physical sickness. Not only was the problem to which I had addressed myself more complicated than I had supposed, it was of a different kind.
>
> (Black 1991, 390)

Black went on to investigate the problem of intransitivity of voting procedures and published a few papers on the subject. However, when he tried to publish a monograph, it was rejected by the publisher. Moreover when he and R. A. Newing sent a paper to the leading American journal *Econometrica* in 1949 they had to wait for 1½ years before receiving the answer:

> I would like to state that if the interrelationship with Arrow's recent monograph could be brought out clearly throughout the paper, I would like very much to recommend your manuscript for publication in *Econometrica.*
>
> (quoted by Coase 1998, xii)

This answer angered Black, who did not want to relate to Arrow's research that had been carried out later that his own. He withdrew the paper but in

1958 published a book-length account of his ideas in a monograph, *The Theory of Committees and Elections*, containing a historical second part mainly dedicated to Dodgson's work on elections.

20.2 Welfare economics

Arrow's research grew out of another tradition, namely welfare economics. Welfare economics, in turn, has its origin in utilitarian ethics developed by Jeremy Bentham (1748–1832), according to whom "it is the greatest happiness of the greatest number that is the measure of right and wrong." To nineteenth- and twentieth-century economists, the optimization of happiness meant the optimization of "utility," which is a function of the social state of each individual or of society at large. The idea of a utility function is marred by two problems: First, it makes sense to assume that each individual can order the possible social states (this is the so-called ordinal point of view) but it is much less clear that the individual can measure the social states in terms of a number (give each of them a price) (the so-called cardinal point of view). Second, it is not at all clear how to compound the individual utility functions to form a societal utility function measuring the total utility of a social state for society at large. Assuming one has formed cardinal utility functions for each individual, one can add them or multiply them or make some other combination of them to form a societal utility function.

In order to circumvent the problems, mathematical economists Abraham Bergson (1914–2003) and Paul A. Samuelson (1915–2009) introduced a general societal welfare function that was determined by some explicitly defined assumptions (axioms). The welfare function was a function of a great number of economic variables and was, in principle, only defined ordinally. However, in order to use mathematical analysis (differential equations, etc.) on the welfare function they usually worked with a real-valued, so-called cardinal indicator whose value is supposed to measure the welfare of society.

The American mathematical economist Kenneth J. Arrow (Figure 20.1) consistently appropriated the ordinal view rejecting the use of cardinal indicators.

> I quickly perceived that the ordinalist viewpoint, which I had fully adopted, implied that the only preference information that could be transmitted across individuals was an ordering. Social welfare could only be an aggregate of orderings.
>
> (Arrow 1983, 3)

He was able to avoid any use of a real-valued indicator by using the abstract mathematical concept of an ordered set. He had learned about this modern concept as a student when he followed a course by the famous Polish logician Alfred Tarski (1911–96). After he graduated with a master's degree in mathematical economics and having joined the war effort he began his Ph.D. studies on a traditional economic subject. However, the problem of the ordinal approach to welfare economics continued to occupy him. From Arrow's point of view, the main problem was reduced to a problem of voting. Indeed, Arrow considered a situation where there is only a finite number of social states to choose from. Each individual will then make an ordered list of his/her preferences corresponding in the welfare context to the individual's utility function, and in the voting context to the individual's ballot. The problem is then from these individual orderings to form a societal ordering of the social states. It is this mapping of the set of the individual ordered lists into the societal ordering that constitute what Arrow called the welfare function.[2] In this way the problem of welfare economics and the theory of voting became fused.

Figure 20.1 Kenneth J. Arrow (1921–2017)
Chuck Painter / Stanford News Service

Continuing along this line of research, Arrow soon (in 1946 or 1947) rediscovered what he called "an unpleasant surprise," namely the Condorcet paradox, and about a year later the solution in the case of single peaked

[2] Be aware that this is a very different object than Bergson's and Samuelson's welfare function.

preferences. He was rather enthusiastic about the latter, but soon realized that Black had already made the discovery earlier.

In 1948 Arrow joined the RAND Corporation where they were trying to develop game theory to analyze military conflict. In a military conflict, the players are countries and thus the question arises as to how one can assign utility functions to them. Applying his ordinal approach and the lesson he had learned from Condorcet's paradox, Arrow soon realized

> that one could not always derive a preference ordering for a nation from the preference orderings of its citizens by using majority voting to compare one alternative with another.
>
> This left open the possibility that there were other ways of aggregating preference orderings to form a social ordering, that is a way of choosing among alternatives that has the property of transitivity. A few weeks of intensive thought made the answer clear.
>
> (Arrow 1986, 50)

The answer was his impossibility theorem, according to which there does not exist a welfare function that preserves transitivity and satisfies some rather innocent-looking requirements. He decided to make this surprising discovery the main subject of his thesis and published it in a paper in 1950 and in an extended version in a monograph the following year (Arrow 1951).

20.3 The Impossibility theorem

The conditions posed by Arrow are of two types. First, he assumed that both the individual's preference ordering and the resulting societal preference ordering are true orderings. Using the symbol R instead of $\geq$, this means that given two alternatives then either xRy or yRx. Moreover, the ordering is supposed to be transitive so that xRy and yRz imply that xRz. Arrow allowed a voter to be indifferent about two or more alternatives in the sense that xRy and yRx.

Second, the welfare function is supposed to satisfy five conditions that he formulated in both mathematical and ordinary language. I shall mention the verbal intuitive formulations.

The *first condition* concerns the individual's possibility to freely choose his or her ordering of the alternatives. He did not assume that all voters could choose any ordering they fancied, but that

> there are at least three among all the alternatives under consideration for which the ordering by any given individual is completely unknown in advance.
>
> (Arrow 1951, 24)

The next two conditions concern how changes in the voters' orderings will influence the societal ordering.

The *second condition* was what Arrow called a "positive association of social and individual values" (Arrow 1951, 25):

> Since we are trying to describe social *welfare* and not some sort of *illfare*, we must assume that the social welfare function is such that the social ordering responds positively to alterations in individual values, or at least not negatively. Hence, if one alternative social state rises or remains still in the orderings of every individual without any other change in those orderings, we expect that it rises, or at least does not fall, in the social ordering.
>
> (Arrow 1951, 25)

In order to understand Arrow's *third condition*, we need to introduce his notion of a social choice function $C(S)$. Here S is a subset of the alternatives and $C(S)$ denotes the winner (or winners) in this set, i.e., an alternative x such that for all y in S, xRy. In terms of this notion his third condition, also called the "independence of irrelevant alternatives," says:

> Just as for a single individual, the choice made by society from any given set of alternatives should be independent of the very existence of alternatives outside the given set.
>
> (Arrow 1950, 337)

The *fourth condition* states:

> The social welfare function is not to be imposed.
>
> (Arrow 1951, 29)

That means that if x and y are two arbitrary alternative social states, the voters should be able to vote in such a way that society prefers x to y. This ensures "that the individuals in our society are free to choose, by varying their values, among the alternatives available" (Arrow 1951, 28). Arrow called condition 4 "the condition of citizen's sovereignty."

> *Condition 5*: The social welfare function is not to be dictatorial (nondictatorship).
>
> (Arrow 1950, 339; 1951, 30)

An individual is called a dictator if the societal ranking will be equal to his or her ranking irrespectively of how the other voters rank the alternatives. The only decisions left to other individuals are the rankings of the alternatives the dictator is indifferent about.

While earlier contributors of the theory of voting had considered one voting method after another, investigating which properties (desirable or unwanted) they have, Arrow turned the question around. He formulated some weak conditions he thought any good voting method must have, and then investigated what one can deduce about such methods. This also reflects the new axiomatic trend in pure mathematics, starting with axioms and deducing consequences from them. However, in Arrow's case the result of the deductions was surprising: Arrow deduced that no voting method satisfied the five conditions: "Thus we have shown that Conditions 1–5 taken together lead to a contradiction" (Arrow 1951, 58).

He formulated his result as follows:

> Possibility Theorem. If there are at least three alternatives among which the members of society are free to order in any way, then every social welfare function satisfying Conditions 2 and 3 and yielding a social ordering satisfying Axiom I and II must be either imposed or dictatorial.
>
> (Arrow 1950, 342)

We notice that at first Arrow called his theorem a possibility theorem. He later explained that this name was due to his superior at the Cowles Commission, Tjalling Charles Koopmans (1910–85), who opposed the name impossibility theorem because he "dislike(d) the feeling that things could not happen or change" (Arrow 2014, 36). During the 1970s the theorem changed its name to Arrow's impossibility theorem.

In 1972 Arrow was awarded the Nobel Memorial Prize in Economic Sciences for his impossibility theorem.[3] The reason was not the difficulty of the proof; in fact, compared to other recent impossibility proofs of theorems like Fermat's last theorem or Gödel's theorems, it is short (about 10 pages long)

[3] Other contributions to economics were mentioned but the impossibility theorem was the primary reason.

and relatively simple. The reason for its fame is to be found in the surprising conclusion of the theorem and the new mathematical methods that Arrow imported into economics.

The importance of Arrow's impossibility theorem was debated from the start. Many traditional economists like Bergson, Samuelson, and their followers downplayed its importance while others praised it for its decisive novelty. One of Arrow's most important disciples called Arrow's discovery a big bang:

> The big bang that characterized the beginning [of modern social choice theory] took the form of an "impossibility theorem", viz. Arrow's (1950, 1951) "General Possibility Theorem". This theorem ... had a profound impact on the way modern social choice theory developed.
>
> (Sen 1986, 1074)

Another social choice theorist, Jerry S. Kelly (1978), likened Arrow's paper and book to other "extraordinary seminal papers" such as Einstein's papers on relativity, Gödel's paper on undecidability, Crick and Watson's paper on DNA structure, and MacLane's paper on category theory that all profoundly changed their area of science.

Also the interpretation of Arrow's impossibility theorem varies greatly. Some interpret it pessimistically as the end of the dream of democratic rule, while others try to get around the impossibility by relaxing some of the conditions. This latter reaction has another constructive counterpart, namely a great industry of new impossibility results in the social sciences. Indeed, it was shown that one can relax or change Arrow's conditions in many ways and still retain an impossibility result. In 1978 Kelly published a book entitled *Arrow Impossibility Theorems* in which he mentioned about 50 variants of Arrow's original theorem and since then many more have been discovered. These are all important contributions to the new theory of social choice that resulted from Arrow's impossibility theorem.

20.4 The Gibbard–Satterthwaite theorem

Among the many impossibility theorems that followed in the wake of Arrow's paradigmatic theorem, one can single out a particularly important group dealing with the problem of manipulating voting procedures. This problem, which has an independent history reaching back to the eighteenth century, is exemplified very clearly by Allan Gibbard (b. 1942) using a so-called Borda count:

> It is well-known that many voting schemes in common use are subject to individual manipulation. Consider a "rank-order" voting scheme: each voter reports his preferences among the alternatives by ranking them on a ballot; first place on a ballot gives an alternative four votes, second place three, third place two and fourth place one. The alternative with the greatest total number of votes wins. Here is a case in which an individual can manipulate the scheme. There are three voters and four alternatives; a voter *a* ranks the alternatives in order *xyzw* on his ballot; voter *b* in order *wxyz*; and voter *c*'s true preference ordering is *wxyz*. If *c* votes honestly, then the winner is his second choice, *x*, with ten points. If *c* pretends that *x* is his last choice by giving his preference ordering as *wyzx*, then *x* gets only eight points, and *c*'s first choice, *w*, wins with nine points. Thus *c* does best to misrepresent his preferences.
>
> (Gibbard 1973, 587)

In most parliamentary election procedures, the voters are faced with the problem of strategic voting. If you favor a candidate who the poles predict will get so few votes that he or she will not be elected, it is usually more favorable for you to vote for a more popular candidate. How can this be avoided?

The answer is that if there are more than three candidates, the problem cannot be avoided. This impossibility was first conjectured in 1961 in a paper by Michael Dummett (1925–2011) and Robin Farquharson (1930–73). The latter had since the early 1950s been using ideas from game theory to analyze voting procedures (see, e.g., Farquarson 1956) and later published a book on it in 1969. In his paper of 1961 with the philosopher Dummett, the conjecture was formulated as follows:

> We cannot assume that each voter's actual strategy will be determined uniquely by his preference scale. This would be to assume that every voter votes "sincerely," whereas it seems unlikely that there is any voting procedure in which it can never be advantageous for any voter to vote "strategically," i.e. non-sincerely.
>
> (Dummett and Farquharson 1961, 34)

After 1961 Farquharson dropped out of the field and Dummett believed that a proof of the conjecture would be too difficult. Thus, the first to prove the conjecture was the mathematically trained philosopher Gibbard. He followed a suggestion of William S. Vickrey (1914–96), who in 1960 had shown that a voting procedure satisfying Arrow's conditions 2 and 3 would not be manipulable. Vickrey also conjectured that the converse would be true, but admitted "that it is not quite so easy to provide a formal proof for this" (Vickrey 1960, 518). Gibbard provided a proof of the converse in 1971 and thus established the theorem: "Every voting scheme with at least three outcomes is either dictatorial or manipulable" (Gibbard 1973, 395).

Before Gibbard's paper was published in 1973, the mathematical economist Mark Allen Satterthwaite (b. 1945) had independently found a proof of the same theorem as a part of his Ph.D. studies and submitted it to the *Journal of Econometric Theory*. In an e-mail to me he explained how he hit upon the result:

> In the summer of 1971 at the time I was first formulating the idea of strategy-proofness I found Farquarson's book, "The Theory of Voting" in the stacks of the University of Wisconsin library. I had concocted a complicated voting rule that I thought might be strategy-proof. After reading parts of his book I quickly constructed a counter-example showing that my procedure was not strategy-proof. I was familiar with Arrow's theorem, and it was at that point I conjectured that no acceptable strategy-proof procedure existed for three or more alternatives. The (inefficient) proof strategy I then used was inspired by the ideas I had learned in a course I had taken in the spring of 1967 during my senior year at Caltech. It was a course on "Algorithms" taught by a young assistant professor of mathematics, Donald Knuth, who (as I am sure you know) later moved to Stanford and became a giant of computer science. (personal e-mail communication from Mark A. Satterthwaite, May 14, 2018)

His proof was a direct proof by induction independent of Arrow's impossibility theorem. During the refereeing process, Satterthwaite was made aware of Gibbard's paper, and so in his published paper he referred to the main theorem as "Theorem 1. (Gibbard–Satterthwaite)," a name it has retained.

Having formulated the main theorem—"The voting procedure *vnm* [described in exact mathematical terms] is strategy-proof if and only if it is dictatorial" (Satterthwaite 1975, 193)—he commented:

> This is formally a possibility theorem, but in substance is that of an impossibility theorem because no committee with democratic ideals will use a dictatorial voting procedure. Such a voting procedure vests all power in one individual, an unacceptable distribution.
>
> (Satterthwaite 1975, 193)

21
Conclusion

In this book we have followed the history of some of the most famous mathematical impossibility theorems. These histories have followed paths that were specific to each theorem. Still, we have detected some patterns or common trends across the various examples. This conclusion shall sum up some of these trends.

21.1 From unimportant non-results to remarkable "solutions"

Today, impossibility results are among the most famous theorems of mathematics. However, we have seen that in earlier times an impossibility result was often considered as a rather uninteresting lack of result and as such second in importance to a positive result. Wallis explicitly voiced such opinions, and we have met several more implicit indications of such a negative assessment of impossibility results: The first true impossibility result, the incommensurability of the side and the diagonal in a square (or a pentagon), was not explicitly addressed in Euclid's elements; Fermat downplayed his impossibility results when he presented his results in number theory; Lagrange displayed no interest in the question of the impossibility of solving the quintic by radicals; Gauss did not publish the impossibility part of his characterization of the constructible regular polygons; Wantzel's proof of the impossibility of the duplication of a cube and the trisection of an angle was forgotten for a century; etc. However, in the first half of the nineteenth century, a number of young mathematicians addressed various impossibility questions: Abel, Galois, and Wantzel dealt with the impossibility of solving polynomial equations by radicals. Their algebraic methods were used by Wantzel himself to show the above-mentioned impossibility of two of the classical construction problems and by Liouville to prove the impossibility of integrating certain functions in elementary form. During the last part of the nineteenth century other important impossibility results were proved such, as the impossibility of the quadrature of a circle by ruler and compass and the impossibility of proving the parallel postulate. These results became almost instantly famous and so in

A History of Mathematical Impossibility. Jesper Lützen, Oxford University Press.
 DOI: 10.1093/oso/9780192867391.003.0021

1900, Hilbert could declare: "the question of the impossibility of certain solutions plays a prominent part [of recent mathematics]." Impossibility theorems had finally obtained full citizenship in the realm of mathematics.

21.2 From meta-statements to mathematical theorems

We have also seen that even in cases where an impossibility question did attract some interest, it happened that it was at first considered as a meta-question *about* mathematics rather than as a proper question within mathematics. Such an asymmetry between positive statements and impossibilities is not so strange, considering that in other areas of life, the solution of a problem has a different nature than an argument of impossibility. Another reason for the different status of positive and negative statements is to be found in the absence of any obvious mathematical strategy for proving impossibilities.

That is quite visible in the first case where an impossibility was considered as having a meta-mathematical nature, namely the classical construction problems. At least from the time of Euclid, mathematicians have developed good strategies for solving construction problems by ruler and compass (and by other means). But it must have been completely unclear to the ancient Greeks how one can prove that a problem cannot be constructed by given means. This is apparent in Pappus' discussion of the problems. He believed that the duplication of a cube and the trisection of an angle were impossible by ruler and compass, but he never pointed to the desirability or even the possibility of a mathematical proof of the impossibility.

21.3 Different types of problems and different types of proofs

Thus, in Greek antiquity it was known how to prove incommensurability using a proof by contradiction. However, it was unclear how to extend this method of proof to the impossibility of the classical construction problems. The reason is the different nature of the impossibility claims. In the first case, it an object in mathematical theory that does not exist; in the latter case it is a construction procedure that does not exist. Methods for handling the latter kind of problems only arose in the seventeenth century in the form of analytic (algebraic) techniques. We have seen that Gregory explicitly emphasized that new analytic methods could be used to establish impossibilities, and we have seen that several mathematicians of the early modern period thought that they had shown the impossibility of the classical construction problems by analytic means. However, new more forceful algebraic methods were required before the impossibility of ruler and compass constructions could be established rigorously.

The new algebraic methods were developed in connection with the problem of finding a solution of the general quintic equation by radicals. This is a problem of the same kind as the solution of the classical problems. It was not the existence of a solution of the equations that was up for discussion. That problem was answered by the fundamental theorem of algebra. The problem was if a particular algorithm or formula could lead to the result. Here it is also remarkable that Lagrange, who contributed many of the central ideas for solving the problem, did not see the possibility of their use in a proof of impossibility. He seems to have considered the problem as a meta-problem.

The independence of the parallel postulate (i.e., its improvability) from other postulates of geometry is yet another type of impossibility result. It is the impossibility of finding a proof that is at stake. Again, when the first mathematicians became convinced that such a proof does not exist, they had no method to prove it. Instead of proving it, they began developing a non-Euclidean geometry based on the negation of the parallel postulate. The consistency of the new non-Euclidean geometry would imply that the parallel postulate could not be proved. Lobachevsky formulated a weak defense for the consistency of the new geometry when he pointed out that it led to meaningful analytic results. But the first true proof of consistency only came about a quarter of a century later, when Beltrami used Gauss' theory of surfaces to construct a model (he called it a real substrate) of non-Euclidean geometry inside the usual Euclidean geometry. An inconsistency in non-Euclidean geometry would turn up in the model as an inconsistency in Euclidean geometry. Thus, if the usual Euclidean geometry is consistent, so is the new non-Euclidean geometry. The method of finding models of one mathematical theory inside another as a means of proving independence (improvability) became (with Hilbert) a standard method. However, we have seen that its original discoverer, Beltrami, at first did not see that his model implied the improvability of the parallel postulate. The method was by no means obvious and its late discovery partly explains why the question of improvability remained a meta-question for so long.

The method of model building can be used to prove the independence of one particular axiom in one particular axiomatic system, and the result is always relative to the consistency of some other system. By this method non-Euclidean geometry could be proved to be consistent relative to Euclidean geometry. Hilbert continued along the same path, using the Cartesian arithmetization of geometry to argue that Euclidean geometry is consistent if the arithmetic of the real numbers is. In order to ground this chain of relative consistency proofs, Hilbert wanted to prove the absolute consistency of arithmetic and show the completeness of the system in the sense that all well-formed statements can be either proved or disproved from the axioms. If accomplished, Hilbert's optimistic dream would have ended all uncertainty in mathematics. However, as we have seen, Gödel showed that such proofs were impossible. And again his proof of the new type of impossibility called for new methods.

21.4 Pure and applied impossibility theorems

Since its origins, mathematics has been applied to different areas of life: Trade, administration, architecture, astronomy, physics, other natural sciences, economics, etc. However, for most of its history impossibility theorems have been a specialty of pure mathematics. That changed in the twentieth century. If one googles "impossibility theorem," Arrow's impossibility theorem gets the most hits. Of course, the primary explanation of the many hits is that it is one of the few impossibility theorems that is explicitly named "impossibility theorem." Nevertheless, it is also indicative for the new trends of impossibilities that an impossibility theorem in applied mathematics and in particular in the social sciences has obtained such a fame. Its unsettling conclusions concerning democratic decision procedures appeal to many political scientists and citizens, including those who may not understand its mathematical proof.

In physics, we have also seen that impossibility theorems have become important. We have considered what is probably the oldest impossibility statement in physics, the impossibility of a perpetual motion machine. In particular we have seen that its status has changed over the course of time, from being an empirical or common sense fact to being a theoretical consequence of the fundamental laws of physics. Finally, we have considered von Neumann's proof that it is impossible to give a deterministic account of quantum mechanics using hidden variables. While the mathematics of the proof is well understood, the interpretation of the physical significance of the result is still debated among historians and philosophers of science.

21.5 Controversies

In fact, the history of mathematical impossibility has been full of controversies of various kinds. Some dealt with questions of priority, for example, Wallis claiming to have proved the non-algebraic nature of the quadrature of a circle before Gregory, and Black claiming to have discovered various results in the theory of voting before Arrow. Some questions of priority have been raised by later historians as, for example, who was the first to prove the impossibility of solving the quintic by radicals, Ruffini or Abel? Other controversies have dealt with mathematics proper. In particular, the seventeenth-century treatment of the classical problems was riddled by discussions about the validity of purported constructions and proofs of impossibility. In many cases, the controversy arose because the mathematical questions, concepts, and methods were not formulated in an unambiguous way. In particular, we noted that the order of the quantifiers caused problems. The concepts were simply too fluid to support rigorous arguments. Other disputes arose because the actors did not agree about what would count as a legitimate solution to a problem: Are neusis constructions allowed in geometry, and are they better or worse than solid constructions using conic sections? Are transcendental curves allowed in geometry? Is the axiom of choice acceptable? Finally, many controversies have dealt with the interpretation of the impossibility theorems: Is Arrow's impossibility theorem a problem for welfare economics? Does von Neumann's theorem about hidden variables rule out a deterministic description of nature? What does Gödel's theorems say about Hilbert's formalistic program? Does Beltrami's model of non-Euclidean geometry prove that all geometry is Euclidean after all, or does it show that non-Euclidean geometry is just as good as Euclidean geometry?

21.6 Impossibility as a creative force

Impossibilities in ordinary life are often considered challenges. How can we get around the impossibility? In mathematics, one may at first think that it would be different. When an impossibility is proven by mathematical means, one cannot deny it except if it turns out that the proof is wrong, and that happens very rarely. Yet, mathematical impossibilities usually do not function as roadblocks but as challenges. The reason is that a mathematical impossibility is often formulated as follows: It is impossible to accomplish a particular task by some precisely specified means. The trick is then to change the means allowed for the solution. In fact, in many cases the way around an impossibility is already available before the impossibility has been properly proved. For example, the classical construction problems had been solved by other curves and instruments in ancient Greece long before the constructions with ruler and compass were proved impossible. Numerical procedures for solving polynomial equations were well known before it was shown that they could not be solved in general by radicals.

But even in cases where the impossibility at first appears as absolute, mathematicians have often been able to circumvent the impossibility. The invention of complex numbers is a splendid example of the method of ideal elements, as Hilbert named it. It allows us to get around the seemingly absolute impossibility of finding a number whose square is -1. Another example is the problem of finding an intersection point of two parallel lines. It does not exist in Euclidean geometry, but one may add suitable points at infinity, in such a way, that they can function as the intersection points of parallel lines. This was done in projective geometry in the nineteenth century. Finally, let me mention an example from modern analysis: In both pure and applied mathematics one often wants to differentiate non-differentiable functions. This is impossible in classical rigorous analysis, but when Laurent Schwartz (*c.*1950) extended the field of functions to so-called distributions it became possible.

In this way, impossibilities have functioned as a creative force in mathematics, resulting in new theories extending older, narrower ones. This mechanism accounts for much of the richness of modern mathematics.

Recommended Supplementary Reading

1. *Introduction.* There are many good general histories of mathematics. Katz (2009) is probably the most complete up-to-date survey. For most of the subjects discussed in the following chapters, Katz gives a fine introduction to the general mathematical context. General treatments of the question of impossibility in mathematics and elsewhere can be found in Richeson (2019), Stillwell (2018), Barrow (1998), and Davis and Park (1987)
2. *Prehistory: Recorded and Non-recorded Impossibilities.* Egyptian mathematics is covered in Gillings (1972) and Imhausen (2016), Babylonian mathematics in Neugebauer (1951) and Van der Waerden (1961), and the social history (Robson 2008). Many Babylonian mathematical texts are published and translated in Neugebauer and Sachs (1945). A more up-to-date account can be found in Høyrup (2002). Katz (2007) presents and analyses sources from many ancient and medieval non-European cultures.
3. *The First Impossibility Proof: Incommensurability.* Van der Waerden (1961) and Heath (1921) are two classical histories of ancient Greek mathematics. A broader more contextualized and up-to-date account can be found in Cuomo (2001).
4. *The Classical Problems in Antiquity: Constructions and Positive Theorems.* Knorr (1986) is the most extensive investigation of Greek constructions of problems. Many of the primary sources are translated in Bulmer-Thomas (1939). The quadrature of a circle is investigated in Hobson (1913).
5. *The Classical Problems: The Impossibility Question.* The subject of this chapter is discussed in Knorr (1986). A translation into French of *Pappus Collection* can be found in (Pappus of Alexandria 1933).
6. *Diorisms and Conclusions about the Greeks and the Medieval Arabs.* See Van der Waerden (1961) and Heath (1921). For a fine account of Arab mathematics see Berggren (2016).
7. *Cube Duplication and Angle Trisection in the Seventeenth and Eighteenth Centuries.* A deeper analysis can be found in Lützen (2010).
8. *Circle Quadrature in the Seventeenth century.* This is the subject of a great book by Davide Crippa (2019). See also Lützen (2014).
9. *Circle Quadrature in the Eighteenth Century.* To my knowledge, there is no secondary source covering this material.
10. *Impossible Equations Made Possible: The Complex Numbers.* The history of algebra between Cardano and Lagrange is covered in Stedall (2011). Leo Corry (2015) has given a semi-popular and very readable account of the development of the number concept. The history of the infinitesimal calculus is the subject of many good monographs. The great lines can be found in Katz (2009).
11. *Euler and the Bridges of Königsberg.* On the history of graph theory see Biggs et al. (1976).
12. *The Insolvability of the Quintic by Radicals.* There are many accounts of this central theme in the history of algebra and its continuation in Galois theory, e.g., Edwards (1984) and Kiernan (1971).

13. *Constructions with Ruler and Compass: The Final Impossibility Proofs.* For Wantzel's proof of the impossibility of the duplication of a cube and the trisection of an angle, see Lützen (2009). For the quadrature of a circle and the transcendence of π see Beckmann (1970) and Berggren et al. (2004).
14. *Impossible Integrals.* The subject is discussed in Lützen (1990, Chapter IX).
15. *Impossibility of Proving the Parallel Postulate.* There are several good books on the history of non-Euclidean geometry including Bonola (1955), Rosenfeld (1988), Gray (1989) and Toretti (1984). Stillwell (1996) contains English translations of important papers on the subject. Voelke (2005) and Volkert (2013) shed light on the reception of non-Euclidean geometry around 1870.
16. *Hilbert and Impossible Problems.* Yandell (2003) and Gray (2000) are both commendable accounts of the history of the Hilbert problems.
17. *Hilbert and Gödel on Axiomatization and Incompleteness.* Nagel and Newman (1958) is a highly recommended popular explanation and history of Gödel's theorems. Dawson (1984) is a fine discussion of the reception of Gödel's theorems. More rigorous introductions can be found in Franzén (2005) and Stillwell (2010).
18. *Fermat's Last Theorem.* Harold Edwards has written a good and detailed history of the advances made before 1977 when his book was published. After Wiles' proof, Singh (1997) published a delightful popular account of the history of the proof, with special emphasis on Wiles' approach.
19. *Impossibility in Physics.* Philosophically inclined readers will enjoy Barrow (1998). It deals with impossibilities in physics and in other sciences.
20. *Arrow's Impossibility Theorem.* For a more extended analysis of the emergence of Arrow's impossibility theorem see Lützen (2019).

References

Abel, Niels Henrik (1826) Beweis der Unmöglichkeit algebraische Gleichungen von höheren Graden als dem vierten allgemein aufzulösen. *Journal für die reine und angewandte Mathematik* 1. 1826, 65–84.

Abel, Niels Henrik (1828) Sur la résolution algébrique des équations. First published in (Abel 1839). Page references are to (Abel 1881), vol II, 217–43.

Abel, Niels Henrik (1829) Précis d'une théorie des fonctions elliptiques. *Journal für die reine und angewandte Mathematik* 4, 236–77. In (Abel 1881) vol I, 518–617.

Abel, Niels Henrik (1830) Lettre à Legendre (written November 25, 1828). *Journal für die reine und angewandte Mathematik* 6, 73–80. In (Abel 1881) vol II, 271–9.

Abel, Niels Henrik (1839) *Oeuvres complètes, avec des notes et développements.* Edited by B. M. Holmboe. Grøndahl, Christiania.

Abel, Niels Henrik (1881) *Œuvres complètes de Niels Henrik Abel, Nouvelle Édition.* Ed L. Sylow and S. Lie, 2 vols. Grøndahl, Christiania.

Alexandrov, Pavel Sergeyevich (Ed.) (1971) *Die Hilbertschen Probleme.* Akademische Verlagsgesellschaft, Leipzig.

Al-Ṭūs, Sharaf al-Dīn (1985) Œuvres mathématiques. Edited and translated by R. Rashed, 2 vols. Les Beles Lettres, Paris.

Anonymous (1722) Sur la rectification indéfinie de Arcs de Cercle. Histoire de l'Académie Royale des Sciences, pour l'Année 1720, avec les Mémoires de Mathematique et de Phisique pour la même Année. 55–64. Paris.

Archimedes (1897) *The Works of Archimedes.* Edited by Thomas L. Heath. C. J. Clay and Sons, London.

Arnol'd, Vladimir I. (1990) *Huygens and Barrow, Newton and Hooke.* Birkhäuser, Basel.

Arnol'd, Vladimir I. and Vasil'ev, Valerii Aleksandrovich (1989) Newton's Principia read 300 years later. *Notices of the American Mathematical Society* 36, 1148–54.

Arrow, Kenneth J. (1950) A difficulty in the concept of social welfare. *Journal of Political Economy* 58, 328–46.

Arrow, Kenneth J. (1951) *Social Choice and Individual Values.* Wiley, New York.

Arrow, Kenneth J. (1983) Headnote to Chapter 1 (On his way to the impossibility theorem). *Collected Papers of Kenneth J. Arrow*, vol 1: *Social Choice and Justice*, 1–4. Belknap Press, Cambridge, MA.

Arrow, Kenneth J. (1986) Kenneth J. Arrow. In William Breit and Roger W. Spencer (Eds), *Lives of the Laureates. Seven Nobel Economists*, 43–58. MIT Press, Cambridge, MA.

Arrow, Kenneth J. (2014) Commentary. In Maskin and Sen (2014), 36–8.

Arrow, Kennth J. and Intriligator, Michael D. (1986) *Handbook of Mathematical Economics*, Vol. III. North-Holland, Amsterdam.

Ayoub, Raymond G. (1980) Paolo Ruffini's contributions to the quintic. *Archive for History of Exact Sciences*, 23, 253–77.

Barrow, John D. (1998) *Impossibility: The Limits of Science and the Science of Limits.* Oxford University Press, Oxford.

Beckmann, Petr (1970) *A History of Pi.* Golem Press, Bolder, CO.

Beltrami, Eugenio (1868) Saggio di Interpretazione della Geometria non-Euclidea. Giornale di Mathematiche VI, 284–312. English translation in Stillwell (1996), 1–34.

Berggren, Lenard (1996) Al-Kūihī's "Filling a Lacuna in Book II of Archimedes" in the Version of Naṣīr al-Dīn al-Ṭūsī. *Centaurus*, 38, 140–70.

Berggren, Lenard (2016) *Episodes in the Mathematics of Medieval Islam*, 2nd edn. Springer, New York.

Berggren, Lennard, Borwein, Jonathan M., and Borwein, Peter B. (2004) *Pi, a Source Book*, 3rd edn. Springer, New York.

Berkeley, George (1734) *The Analyst.* Tonson, London. In Smith (1959), 627–34.

Biggs, Norman L., Lloyd, E. Keith, and Wilson, Robin J. (1976) *Graph Theory 1736–1936.* Clarendon Press. Oxford.

Black, Duncan (1958) *The Theory of Committees and Elections.* Cambridge University Press, Cambridge, UK.

Black, Duncan (1991) Arrow's work and the normative theory of committees. *Journal of Theoretical Politics*, 3, 259–76. In Black (1998), 387–405.

Black, Duncan (1998) *The Theory of Committees and Elections by Duncan Black and Committee Decisions with Complementary Valuation by Duncan Black and R. A.* Newing, revised 2nd edn. Edited by Iain McLean, Alistair McMillan, and Burt L. Monroe. Kluwer, Boston.

Blåsjö, Viktor (2016) In defence of geometrical algebra. *Archive for History of Exact Sciences* 70, 325–59.

Blåsjö, Viktor (2017) *Transcendental Curves in the Leibnizian Calculus.* Elsevier, Amsterdam.

Bonola, Roberto (1955) *Non-Euclidean Geometry.* Dover, New York.

Bos, Henk J. M. (1974) Differentials, higher-order differentials and the derivative in the Leibnizian calculus. *Archive for History of Exact Sciences*, 14, 1–90.

Bos, Henk J. M. (1984) Arguments on motivation in the rise and decline of a mathematical theory; the "construction of equations" 1637– ca. 1750. *Archive for History of Exact Sciences*, 30, 331–80.

Bos, Henk J. M. (2001) *Redefining Geometrical Exactness: Descartes' Transformation of the Early Modern Concept of Construction.* Springer-Verlag, New York.

Bulmer-Thomas, Ivor (1939) *Selections Illustrating the History of Greek Mathematics.* Harvard University Press, Cambridge, MA.

Cajori, Florian (1918) Pierre Laurent Wantzel. *Bulletin of the American Mathematical Society* 24, 339–47 (read before the AMS 1917).

Cardano, Girolamo (1968) *The Great Art or the Rules of Algebra.* Translated and edited by T. Richard Witmer. MIT Press, Cambridge, MA.

Cauchy, Augustin Louis (1813) Mémoire sur le nombre des valeurs qu'une fonction peut Acquérir lorsq'on y permute de toutes les manières possibles les quantités qu'elle renferme. *Journal de l'École Polytechnique* XVII Cahier, X, 1–20. Œuvres de Cauchy Série II, tome 1, 64–82.

Chase, Arnold B. (1967) *The Rhind Mathematical Papyrus.* National Council of Teachers of Mathematics, Reston, VA.

Chemla, Karine (Ed.) (2012) *The History of Mathematical Proof In Ancient Traditions.* Cambridge University Press, Cambridge, UK.

Coase, Ronald H. (1998) Foreword to Black (1998), ix–xv.

Condorcet, Marie Jean A.Nicolas de (1775/8) Motivating declaration from the Académie des Sciences with no title and no author. Histoire de l'Académie Royale des Sciences for the year 1775, 61–66.

Condorcet, Marie Jean A.Nicolas de (1785) *Essai sur l'application de l'analyse à la probabilité des décisions rendues à la pluralité des voix.* L'imprimerie Royale, Paris.
Coolidge, Julian Lowell (1940) *A History of Geometrical Methods.* Clarendon Press, Oxford.
Corry, Leo (2015) *A Brief History of Numbers.* Oxford University Press, Oxford.
Costabel, Pierre (2008) Lagny, Thomas Fantet De. Dictionary of Scientific Biography (New York 1970–1990). Complete Dictionary of Scientific Biography, http://www.encyclopedia.com/doc/1G2-2830902425.html (July 18. 2012).
Crippa, Davide (2019) *The Impossibility of Squaring the Circle in the 17th Century.* Birkhäuser, Cham.
Cuomo, Serafina (2001) *Ancient Mathematics.* Routledge, New York.
D'Alembert, Jean le Rond (1751–66) *Encyclopédie ou Dictionnaire raisonné des sciences, des arts et des métiers.* Briasson, Paris.
D'Alembert, Jean le Rond (1765) Quadrature du cercle. In *L'Encyclopédie ou Dictionnaire raisonné des sciences, des arts et des métiers,* Vol. 13. Samuel Faulche, Neufchastel.
D'Alembert, Jean le Rond (1768) Sur un autre paradoxe. Opuscules Mathématiques ou Mémoires sur differens sujets de Géométrie, de Méchanique, d'Optique, d'Astronomie &c.,4. Briason, Paris.
Davis, Philip J., and Park, David (Eds) (1987) *No Way: The Nature of the Impossible.* W. H. Freeman, New York.
Dawson, John W. (1984) The reception of Gödel's incompleteness theorems. PSA: Proceedings of the Biennial Meeting of the Philosophy of Science Association 1984, 253–71.
De Risi, Vincenzo (2016) *Leibniz on the Parallel Postulate and the Foundations of Geometry: The Unpublished Manuscripts.* Springer, Cham.
Dehn, Max and Hellinger, Ernst (1939) On James Gregory's Vera Quadratura. In *James Gregory Tercentenary Memorial Volume.* Edited by Herbert Westren Turnbull. Bell and sons, London.
Delambre, Jean-Baptiste Joseph (1810) *Rapport historique sur les progrès des sciences mathématiques depuis 1789.* Imprimérie Inpériale, Paris. Reprinted 2011 by Cambridge University Press, Cambridge, UK.
Descartes, René (1637) La Géométrie. In *Discours de la Méthode,* 297–413. Leiden . Facsimile in *The Geometry of René Descartes.* Edited and translated by D. E. Smith and M. L Latham. Dover, New York, 1954.
Dieks, Dennis (2017) Von Neumann's impossibility proof: Mathematics in the service of rhetorics. *Studies in History and Philosophy of Modern Physics,* 60, 136–48.
Diophantus (1621) *Diophanti Alexandrini Arithmeticorum libri sex, et numeris multangulis liber unus.* Edited by Claude Gaspard Bachet de Mézirac. Jerome Drouart, Paris.
Diophantus/Fermat (1670) *Diophanti Alexandrini Arithmeticorum libri sex, et numeris multangulis liber unus Ed. Samnel de Fermat avec les commentaires de C.G.* Bachetet les observations de P. de Fermat et l'Inventum novum. B. Bosc, Toulouse.
Du Sautoy, Marcus (2016) *What We Cannot Know: From Consciousness to the Cosmos, the Cutting Edge of Science Explained.* Fourth Estate. London
Dummett, Michael, and Farquharson, Robin (1961) Stability in voting. *Econometrica* 29, 33–43.
Edwards, Harold M. (1977) *Fermat's Last Theorem.* Springer, New York.
Edwards, Harold M. (1984) *Galois Theory.* Springer, New York.
Euclid (1956) The Thirteen Books of the Elements. Translated with introduction and commentary by Sir Thomas L. Heath, Vols 1–3. Dover, New York. Another easily accessible and accurate annotated translation of the *Elements* can be found on the Internet: https://mathcs.clarku.edu/~djoyce/elements/elements.html

Euler, Leonhard (1736) Solutio problematis ad geometriam situs pertinentis. Comment. Acad. Sci. U. Petropolitana 8, 128–40. Translated in Briggs et al. (1976), 3–8.

Euler, Leonhard (1748) *Introductio in Analysin Infinitorum* (2 vols). Edited by Lausanne Bousquet. Euler's Opera Omnia (1) 8–9.

Farquharson, Robin (1956) Straightforwardness in Voting Procedures. Oxford Economic Papers, New series vol 8, 80–9.

Farquharson, Robin (1969) *Theory of Voting*. Yale University Press, New Haven.

Ferreirós, José (2008) The Crisis in the Foundations of Mathematics. Section II.7 of *Princeton Companion to Mathematics*. Princeton University Press, Princeton.

Floyd, Juliet (1995) On Saying What You Really Want to Say: Wittgenstein, Gödel and the Trisection of the Angle. In Jaakko Hintikka (Ed.) *From Dedekind to Gödel*, 373–425. Kluwer, Dordrecht.

Franzén, Torkel (2005) *Gödel's Theorem: An Incomplete Guide to Its Use and Abuse*. CRC Press, Boca Raton.

Galilei, Galileo (1638) *Discorsi e dimostrazioni matematiche, intorno à due nuove scienze*. Elsevier, Leiden. Page references to the English translation: (1954) Dialogues Concerning Two New Sciences. Translated by Crew and de Salvio. Dover, New York.

Gauss, Carl Friedrich (1816) Letter to Heinrich Olbers (21 Mar. 1816). Quoted in G. Waldo Dunnington, *Carl Friedrich Gauss: Titan of Science*. Mathematical Association of America (2004), 413.

Gauss, Carl Friedrich (1828) Disquisitiones generals circa superficies curvas. Commentationes societatis regiae scientiarum Gottingensis recentiores, VI, 99–146. English translation in Karl Friedrich Gauss (1965) General Investigations of Curved Surfaces. Translated by A. Hiltebreitel and J. Morehead. Raven Press, Hewlet New York.

Gauss, Carl Friedrich (1801) *Disquisitiones Arithmeticae*. Braunschweig. English translation by Arthur A. Clarke. Yale University Press, New Haven, 1966.

Gibbard, Allan (1973) Manipulation of voting schemes: A general result. *Econometrica* 14, 587–601.

Gilain, Christian (1988) Condorcet et le calcul integral. In R. Rached(Ed.), Science à l'époque de la Révolution Française, Recherches historiques, 87–150. Blanchard, Paris.

Gillings, Richard J. (1972) *Mathematics in the Time of the Pharaos*. MIT Press, New York.

Girard, Albert (1629) Invention Nouvelle en l'Algèbre. Guillaume Iansson Blaeuw, Amsterdam.

Gödel, Kurt (1931) Über formal unentscheidbare Sätze der Principia Mathematica und verwandter Systeme, I. *Monatshefte für Mathematik und Physik* 38, 173–98.

Gödel, Kurt (1933) The present situation in the foundations of mathematics. Lecture delivered at the meeting of the Mathematical Association of America in Cambridge, Massachusetts, December 29–30, 1933, in Kurt Gödel (1995) Collected Works, Vol. III, 45–53. Edited by Solomon Feferman, John W. Dawson, Jr., Warren Goldfarb, Charles Parsons, and Rober M. Solovay. Oxford University Press, Oxford.

Goldstein, Catherine (1995) *Un théorème de Fermat et ses lecteurs*. Presses Universitaires de Vincennes, Saint-Denis.

Gray, Jeremy J. (1989) *Ideas of Space: Euclidean, Non-Euclidean and Relativistic*. Clarendon Press, Oxford.

Gray, Jeremy J. (2000) *The Hilbert Challenge*. Oxford University Press, Oxford.

Gregory, James (1668a) *Exercitationes geometricae*. London. Excerpts are reprinted in Huygens (1895), 313–21.

Gregory, James (1668b) Mr. Gregories Answer. Philosophical Transactions 37, July 13. Page references to reprint in Huygens (1895), 240–3.

Gregory, James (1667) Vera circuli et hyperbolae quadraturae. Padua. Reprinted as an appendix to Gregory (1668a). Page references are to the reprint in Huygens (1724), 405–62.

Gregory, James (1939) *Tercentenary Memorial Volume.* Edited by Herbert W. Turnbull, Bell & Sons, London.

Guicciardini, Niccolo (2009) *Isaac Newton on Mathematical Certainty and Method.* MIT Press, Cambridge, MA.

Hamilton, William R (1839) On the argument of Abel, respecting the impossibility of expressing a root of any general equation above the fourth degree, by any finite combination of radicals and rational functions. *Transactions of the Royal Irish Academy* 18, 171–259.

Hartshorne, Robin (2000) *Geometry: Euclid and Beyond.* Springer-Verlag, New York.

Hartshorne, Robin (2007) On the impossibility of classical construction problems. Talk given at a meeting in Luminy, April 16–20, 2007.

Heath, Thomas L. (1921) *A History of Greek Mathematics.* Clarendon Press, Oxford.

Heijenoort, Jan van (1981) *From Frege to Gödel. A source Book in Mathematical Logic, 1879–1931.* Harvard University Press, Cambridge, MA.

Heinrich, Georg (1901) James Gregorys "Vera circuli et hyperbolae quadrature." *Bibliotheca Mathematica* 2(3), 77–85.

Hermite, Charles (1873a) Sur la fonction exponentielle. *Comptes rendus hebdomadaires des séances de l'Académie des Sciences* 77, 18–24; 74–79; 226–233; 285–295.

Hermite, Charles (1873b) Extrait d'une lettre de Mr. Ch. Hermite à Mr. Borchardt . *Journal für die reine und angewandte Mathematik*, 76, 342–4. Hilbert, David (1893) Ueber die Transcendenz der Zahlen e und π. *Mathematische Annalen* 43, 216–19.

Hilbert, David (1899/1992) *Grundlagen der Geometrie*, 1st edn. Teubner, Leipzig. 10th edn 1968. Teubner, Stuttgart. Page references to the English translation of the 10th edition: Foundations of Geometry. Open Court, La Salle, IL, 1992.

Hilbert, David (1900) Mathematische Probleme. Archiv für Mathematik und Physik, 3. Reihe 1, 1901, 44–63, 213–37. Gesammelte Abhandlungen, 3, 290–329. Page references to English translation in Gray (2000, 240–82).

Hilbert, David (1930) *David Hilbert's Radio Address.* Ed. and English translation J. T. Smith. Available at http://math.sfsu.edu/smith/Documents/HilbertRadio/HilbertRadio.pdf. Contains a link to a recording of Hilbert's radio address.

Hilbert, David and Bernays, Paul (1939) *Die Grundlagen der Mathematik.* Springer, Berlin.

Hobson, Ernest W. (1913) Squaring the Circle: A History of the Problem. In *Squaring the Circle and Other Monographs*, 1–57, ed. H. P. Hudson et al. Chelsea, New York.

Hogendijk, Jan (1981) How trisections of the angle were transmitted from Greek to Islamic geometry. *Historia Mathematica* 8, 417–438.

Hogendijk, Jan (1984) Greek and Arabic constructions of the regular heptagon. *Archive for History of Exact Sciences*, 30, 197–330.

Hogendijk, Jan (1985) *Ibn al-Haytham's Completion of the Conics.* Springer, New York.

Hogendijk, Jan (1989) Sharaf al-Dīn al-Tūsī on the number of positive Roots of cubic Equations. *Historia Mathematica*, 16, 69–85.

Hogendijk, Jan (1996) *Al-Sijzī's Treatise on Geometrical Problem Solving.* Fatemi Publishing Company, Teheran.

Hogendijk, Jan (2010) The scholar and the fencing master: The exchanges between Joseph Justus Scaliger and Ludolph van Ceulen on the circle quadrature (1594–1596). *Historia Mathematica*, 37, 345–75.

Høyrup, Jens (2002) *Length, Widths, Surfaces: A Portrait of Old Babylonian Algebra and Its Kin*. Springer, New York.

Huygens, Christiaan (1668a) Examen de Vera Circuli & Hyperboles Quadratura in propriâ sua proportionis specie inventa & demonstrata à Jacobo Gregorio Scoto, in 4°. Patavii. Journal des Sçavans, July 2. Page references to reprint in Huygens (1895), 228–30.

Huygens, Christiaan (1668b) Letter to J. Gallois. Journal des Sçavans, November 12. Page references to reprint in Huygens (1895), 272–6.

Huygens, Christiaan (1724) *Opera Varia*. Lugdunis Batavorum, Bibliopolas.

Huygens, Christiaan (1895) *Oeuvres Complètes tome 6, Correspondance 1666-1669*. Nijhoff, La Haye.

Huygens, Christiaan (1940) *Oeuvres Complètes, Tome20, Musique et mathématique de 1666 à 1695*. Nijhoff, La Haye.

Imhausen, Annette (2016) *Mathematics in Ancient Egypt: A Contextual History*. Princeton University Press, Princeton, NJ.

Jacob, Marie (2005) Interdire la quadrature du cercle à l'Académie: Une décision autoritaire des lumières? *Révue d'Histoire des Mathématiques*, 11, 89–139.

Jesseph, Douglas M. (1999) *Squaring the Circle: The War between Hobbes and Wallis*. University of Chicago Press, Chicago.

Katz, Victor J. (2007) The Mathematics of Egypt, Mesopotamia, China, India and Islam. Princeton University Press, Princeton, NJ.

Katz, Victor J. (2009) *A History of Mathematics*, 3rd edn. Addison-Wesley, Boston.

Kelly, Jerry S. (1978) *Arrow Impossibility Theorems*. Academic Press, New York.

Kiernan, B. Melvin (1971) The development of Galois theory from Lagrange to Artin. *Archive for History of Exact Sciences*, 8, 40–154.

Klein, Felix (1895) *Vorträge über ausgewählte Fragen der Elementargeometrie*. Teubner, Leipzig. English translation: Famous Problems of Elementary Geometry, Ginn and Co. New York, 1897.

Knorr, Wilbur (1986) *The Ancient Tradition of Geometric Problems*. Birkhäuser, Boston.

Königsberger, Leo (1869) Berichtigung eines Satzes von Abel, *Mathematische Annalen* 1, 168–9.

Lacki, Jan (2000) The early axiomatizations of quantum mechanics: Jordan, von Neuman and the continuation of Hilbert's program. *Archive for History of Exact Sciences*, 54, 279–318.

Lagny, Thomas Fantet De (1729) Troisième Mémoire sur la goniometrie purement analytique. Histoire de l'Académie Royale des Sciences, anné 1727, avec les Mémoires de Mathematique et de Phisique pour la même Année, 120–30.

Lagrange, Joseph Louis (1770–1) Réflexions sur la Résolution algébrique des équations. *Nouveaux Mémoires de l'Académie royale des Sciences et Belles-Letters de Berlin*. Page references to Lagrange's Oeuvres, vol., 205–421.

Lakatos, Imre (1976) *Proofs and refutations. The logic of mathematical discovery*. Edited by J. Worrall and E. Zahar. Cambridge University Press, Cambridge, UK.

Lambert, Johann Heinrich (1770) *Beyträge zum Gebrauche der Mathematik und deren Anwendung II*. Verlag der Buchhandlung der Realschule, Berlin. http://www.kuttaka.org/~JHL/L1770a.html.

Lambert, Johann Heinrich (1767, publ. 1769) Solution générale et absolue du problème de trois corps moyennant des suites infinies. Histoires et Mémoires de l'Académie Royale des Sciences et Belles-Lettres Berlin, 23, 353–64.

Lambert, Johann Heinrich (1768) Mémoire sur quelques propriétés remarquables des quantités transcendantes circulaires et logarithmiques. Mémoires de l'Académie royale des sciences de Berlin 1761, 265–322. In Lennart Berggren, Jonathan M. Borwein, Peter B. Borwein (Eds.), *Pi, a Source Book*, 3rd edn, 129–40. Springer-Verlag, New York, 2004. Page numbers refer to https://hal.archives-ouvertes.fr/hal-02984214/document.

Laplace, Pierre-Simon (1812) *Théorie analytique de probabilité*. Œuvres de Laplace, vol. 12. Courcier, Paris.

Lapparent, A. de (1895) Pierre-Laurent Wantzel (1814–1848) . *École Polytechnique: Livre du Centenaire, 1794–1894*, vol. 1, 133–5. English translation at <http://www-history.mcs.standrews.ac.uk/Extras/De_Lapparent_Wantzel.html>.

Laubenbacher, Reinhard, and Pengelley, David (2010) "Voici ce que j'ai trouvé:" Sophie Germain's grand plan to prove Fermat's Last Theorem. *Historia Mathematica*, 37, 641–92. https://arxiv.org/pdf/0801.1809.pdf

Leibniz, Gottfried Wilhelm (1684) De dimensionibus figurarum inveniendis. Acta Eruditorum *1684*. Page references to reprint in Leibniz (1858), 123–7.

Leibniz, Gottfried Wilhelm (1858) Leibnizens gesammelte Werke. Ed. G. H. Pertz, 3. Folge 5. Band Has the separate title: *Leibnizens mathematische Schriften*, ed C. I. Gerhardt, 2. Abt. Band I. Verlag von H. W. Smidt, Halle.

Leibniz, Gottfried Wilhelm (1675/6) *De quadratura arithmethic circuli ellipseos et hyperbolae cujus corollarium est trigonometria sine tabulis*. Edited by Eberhard Knobloch (1993). German translation by O. Hamborg. Abhandlungen der Akademie der Wissenschaften in Göttingen, 2007.

L'Hospital, Guillaume François Antoine Marquis de (1720) *Traité analytique des sections coniques et de leur usage pour la resolution des equations dans les problemes tant déterminez qu'indeterminez*. Montallent, Paris.

Lindemann, Ferdinand (1882) Ueber die Zahl π. *Mathematische Annalen*, 20, 213–25.

Liouville, Joseph (1833a) Premier (Second) Mémoire sur la détermination des intégrales dont la valeur est algébrique. *Journal de l'École Polytechnique* 14(22), 124–48, 149–93. Mémoires des Savans Etrangers de lAcadémie des Sciences de Paris 5 (1838), 76–102, 103–51.

Liouville, Joseph (1833b) Note sur la détermination des intégrales dont la valeur est algébrique. *Journal für die reine und angevandte Mathematik*, 10, 347–59.

Liouville, Joseph (1834) Sur les transcendantes elliptiques de première et de seconde espèce, considérées comme fonctions de leurs longitude. *Journal de l'École Polytechnique*, 14(23), 37–83.

Liouville, Joseph (1835) Mémoire sur l'intégration d'une classe de fonctions transcendantes. *Journal für die reine und angevandte Mathematik*, 13, 93–118.

Liouville, Joseph (1837/8) Mémoire sur la classification des transcendantes et sur l'impossibilité d'exprimer les racines de certaines équations en fonction explicite des coefficients. *Journal de Mathématiques pures et appliquées* 2 (1837), 56–105; 3 (1838), 523–47.

Liouville, Joseph (1839) Mémoire sur l'intégration d'une classe d'équations différentielles du second ordre en quantitées finies explicites. *Journal de Mathématiques pures et appliquées* 4, 423–56.

Liouville, Joseph (1840) Mémoires sur les transcendantes elliptiques de première et de second espèce, considérées comme fonctions de leur module. Journal de Mathématiques pures et appliquées 5, 441–64.

Liouville, Joseph (1841) Remarques nouvelles sur l'équation de Riccati. *Journal de Mathématiques pures et appliquées* 6, 1–13.

Liouville, Joseph (1844) Sur des classes très-étendues de quantités dont la valeur n'est ni rationnelle ni même réductible à des irrationnelles algébriques; 2° à un passage du livre des Principes où Newton calcule l'action exercée par une sphère sur un point extérieur. *Comptes Rendus de l'Académie des Sciences de Paris*, 18, 883–5.

Liouville, Joseph (1851) Sur des classes très-étendues de quantités dont la valeur n'est ni rationnelle ni même réductible à des irrationnelles algébriques. *Journal de Mathématiques pures et appliquées*, 16, 133–42.

Loveland, Jeff (2004) Panckoucke and the Circle Squarers. *Eighteenth-Century Studies* 37, No. 2, Spaces of Enlightenment, 215–36.

Lützen, Jesper (1985) *Cirklens kvadratur, vinklens tredeling og terningens fordobling*. Systime, Herning.

Lützen, Jesper (1990) *Joseph Liouville 1809–1882. Master of Pure and Applied Mathematics*. In Studies in the History of Mathematics and Physical Sciences, No. 15. Springer-Verlag, New York.

Lützen, Jesper (2008) Er matematiske umulighedssætninger noget særligt? *Filosofiske Studier: Festskrift tilegnet Docent dr.phil Carl Henrik Koch*. ed. F. Collin and J. R. Flor, Vol. 24, 251–66. Københavns Universitet, København.

Lützen, Jesper (2009) Why was Wantzel overlooked for a century? The changing importance of an impossibility result. *Historia Mathematica*, 36, 374–94.

Lützen, Jesper (2010) The algebra of geometric impossibility: Descartes and Montucla on the impossibility of the duplication of the cube and the trisection of the angle. *Centaurus* 52, 4–37.

Lützen, Jesper (2013) Mathematical Impossibility in History and in the Classroom. *Cuadernos de Investigación y Formación en Educación Matemática*, 8, 165–74.

Lützen, Jesper (2014) 17th century arguments for the impossibility of the indefinite and the definite quadrature of the circle. *Revue d'histoire des mathématiques*, 20, 211–51.

Lützen, Jesper (2019) How mathematical impossibility changed welfare economics: A history of Arrow's impossibility theorem. *Historia Mathematica*, 46, 56–87.

Maanen, Jan A. van (1986) The refutation of Longomontanus' quadrature by John Pell. *Annals of Science* 43, 315–52.

McCurdy, Edward (1906) *Leonardo da Vinci's Note-books*. Charles Scribner's Sons, New York.

Mancosu, Paolo (1996) *Philosophy of Mathematics and Mathematical Practice in the Seventeenth Century*. Oxford University Press, New York.

Mancosu, Paolu (1999) Between Vienna and Berlin: The immediate reception of Gödel's incompleteness theorems. *History and Philosophy of Logic* 20, 33–45.

Maskin, Erik, and Sen, Amartya (2014) *The Arrow Impossibility Theorem*. Columbia University Press, New York.

Mersenne, Marin. (1962) *Correspondance de P. Marin Mersenne*, vol VII. CNRS, Paris.

Montucla, Jean Étienne (1754) *Histoire des recherches sur la Quadrature du Cercle*. Jombert, Paris.

Montucla, Jean Étienne (1758) *Histoire des Mathématiques*, vol. 1. Jombert, Paris.

Montucla, Jean Étienne (1799) *Histoire des Mathématiques. Nouvelle édition considérablement augmentée*, vol. 1. Agasse, Paris.

Montucla, Jean Étienne (1802) *Histoire des Mathématiques. Nouvelle édition considérablement augmentée*, vol. 4. Edited by J. Lalande. Agasse, Paris.

Murdoch, Dugald (1987) *Niels Bohr's Philosophy of Physics*. Cambridge University Press, Cambridge, UK.

Nagel, Ernst, and Newman, James R. (1958) *Gödel's proof*. New York University Press, New York.

Neugebauer, Otto (1951) *The Exact Sciences in Antiquity*. Princeton University Press, Princeton, NJ (new ed. Dover, New York, 1969).

Neugebauer, Otto, and Sachs, Abraham J. (1945) *Mathematical Cuneiform Texts*. American Oriental Series. American Oriental Society and the American Schools of Oriental Research, New Haven, CT.

Newton, Isaac (1687) *Philosophia Naturalis Principia Mathematica*. Reg. Soc., London. Page references to the English translation The Principia by I. Bernard Cohen and Anne Whitman. University of California Press, Berkeley, 1999.

Newton, Isaac (1974) *The Mathematical Papers of Isaac Newton*, Vol 6. Cambridge University Press, Cambridge, UK.

O'Connor, John J., and Robertson, Edmund F. (1998) Paolo Ruffini. MacTutor https://mathshistory.st-andrews.ac.uk/Biographies/Ruffini/ . Accessed June 13, 2020.

Ore, Øystein (1974) *Niels Henrik Abel. Mathematician Extraordinary*, 2nd edn. Chelsea, New York.

Pappus of Alexandria (1933) *Collectio*. Translated into French by Paul Ver Eecke: Pappus d'Alexandrie: La Collection mathématique, I, II. Desclée de Brouwer, Paris.

Pesic, Peter (2001) The validity of Newton's lemma 28. *Historia Mathematica*, 28, 215–19.

Pesic, Peter (2003) *Abel's Proof*. MIT Press, Cambridge, MA.

Petersen, Julius (1871) *Om ligninger, der løses ved kvadratrod, med anvendelse på problemers løsning ved passer og lineal*. Ferslew & Co., Copenhagen.

Petersen, Julius (1877) *De algebraiske Ligningers Theori*. Høst & Søn, Copenhagen, French translation: Théorie des équations algébriques. Høst & Søn, Copenhagen. German translation: Theorie der algebraischen Gleichungen. Høst & Søn, Copenhagen, 1878.

Poisson, Siméon Denis (1833) Rapport sur deux mémoires de M. Liouville ayant pour titre Détermination des intégrales dont la valeur est algébrique. Procès-Verbaux de l'Académie des Sciences de Paris, 211–13.

Pourciau, Bruce (2001) The integrability of ovals: Newton's lemma 28 and its counterexamples. *Archive for History of Exact Sciences*, 55, 479–99.

Ptolemy, Claudius (1998) *Ptolemy's Almagest*. Translated by G. J. Toomer, Princeton University Press, Princeton, NJ.

Raatikainen, Panu (2020) Gödel's incompleteness theorems. The Stanford Encyclopedia of Philosophy (Winter 2020 Edition). Edited by Edward N. Zalta. https://plato.stanford.edu/archives/win2020/entries/goedel-incompleteness/

Rashed, Roshdi (2021) L'algèbre arithmétique au XIIe siècle: Al-Bāhir d'al-Samaw'al. De Gruyter, Berlin.

Richeson, David (2019) *Tales of impossibility* Princeton University Press, Princeton.

Riemann, Bernhard (1868) Über die Hypothesen welche der Geometrie zu Grunde liegen (Habilitationsvortrag 1854). Abhandlungen der Königlichen Gesellschaft zu Göttingen, Mathematische Classe, 13, 1–20. English translation in Smith (1959), 411–25.

Risch, Robert Henry (1970) The solution of the problem of integration in finite terms. *Bulletin of the American Mathematical Society*, 76, 605–8.

Risch, Robert Henry (1976) Implicit elementary integrals. *Proceedings of the American Mathematical Society*, 57, 1–7.

Ritt, Joseph Fels (1948) *Integration in Finite Terms. Liouville's Theory of Elementary Methods*. Columbia University Press, New York.

Robinson, Abraham (1966) *Non-standard Analysis*. North-Holland Publishing, Amsterdam.

Robson, Eleanor (2008) *Mathematics in Anciety Iraq: A Social History*. Princeton University Press, Princton, NJ.

Rosenfeld, Boris and Hogendijk, Jan (2002–3) A mathematical treatise written in the Samarqand Observatory of Ulugh Beg. *Zeitschrift für Geschichte der Arabisch-Islamischen Wissenschaften*, 15, 25–65.

Rosenfeld, Boris A. (1988) *A History of Non-Euclidean Geometry*. Springer Verlag, New York.

Rosenlicht, Maxwell Alexander (1968) Liouville's theorem on function with elementary integrals. Pacific Journal of Mathematics, 24(1), 153–61.

Rosenthal, Franz (1950) Al-Asṭurlâbî and as-Samaw'al on scientific progress. *Osiris*, 9, 555–64.

Rowe, David E. (2015) Historical Events in the Background of Hilbert's Seventh Paris Problem. In David E. Rowe and Wann-Sheng Horng (Eds), *A Delicate Balance: Global Perspectives on Innovation and Tradition in the History of Mathematics A Festschrift in Honor of Joseph W. Dauben*, 211–44. Birkhäuser, Cham.

Ruffini, Paolo (1915) *Opere Matematiche*, 3 vols., Edited by E. Bortolotti. Tipografia matematica di Palermo, Palermo.

Russell, Bertrand (1901) Recent Work on the Principles of Mathematics, *The International Monthly*, 4 (Jul 1901), 83–101. Page numbers to reprinting: Mathematics and the Metaphysicians, in Mysticism and Logic and Other Essays, 74–96. Longmans, Green, London, 1918.

Saccheri, Girolamo (1733) Euclides ab omni Naevo Vindicatus. Montanus, Milan. Edited and translated into English by G. B. Halsted. Open Court, Chicago, 1920.

Sachs, Horst, Stiebitz, Michael, and Wilson, Robin J. (1988) Franz Rosenthal *Journal of Graph Theory*, 12, 133–9.

Saint-Venant, Barré de (1848) Wantzel. *Nouvelles Annales de Mathématiques*, 7, 321–31.

Saint-Vincent, Grégoire de (1647) *Opus geometricum quadraturae circuli et sectionum coni decem libris comprehensum*. Jan van Meurs & Jacob van Meurs, Antwerp.

Satterthwaite, Mark Allen (1975) Strategy-proofness and Arrow's conditions: Existence and correspondence theorems for voting procedures and social welfare functions. *Journal of Economic Theory*, 10, 187–217.

Saurin, Joseph (1722) Démonstration de l'impossibilité de la Quadrature indefinie du Cercle. Avec une maniere simple de trouver une suite de Droites qui aprochent de plus et plus d'un arc de Cercle proposé, tant en dessus qu'en dessous. Histoire de l'Académie Royale des Sciences, anné 1720, avec les Mémoires de Mathematique et de Phisique pour la même Année, 15–19.

Schmidt, Olaf (1980) On Plimpton 322. Pythagorean numbers in Babylonian mathematics. *Centaurus*, 24, 4–13.

Schubring, Gert (2005) *Conflicts Between Generalization, Rigor, and Intuition: Number Concepts Underlying the Development of Analysis in 17th-19th Century France and Germany*. Springer, New York.

Scriba, Christoph. J. (1983) Gregory's converging double sequence: A new look at the controversy between Huygens and Gregory over the "analytical" quadrature of the circle. *Historia Mathematica*, 10, 274–85.

Seevinck, Michiel (2016) Challenging the Gospel: Grete Hermann on von Neumann's no-hidden-variables proof. In Elise Crull and Guido Bacciagalupi (Eds), *Grete Hermann—Between Physics and Philosophy*. Springer, Dordrecht.

Sen, Amartya (1986) Social Choice Theory. In Kennth J. Arrow and Michael D. Intriligator (Eds), *Handbook of Mathematical Economics*, Vol. III, 1073–1181. North-Holland, Amsterdam.

Simanek, Donald E. (2020) Perpetual futility: A short history of the search for perpetual motion. The Museum of Unworkable Devices. Donald Simanek's website, Lock Haven. http://www.lockhaven.edu/~dsimanek/museum/people/people.htm. Acessed November 26, 2020.

Singh, Simon (1997) *Fermat's Last Theorem*. Fourth Estate, London.

Skau, Christian (1990). Gjensyn med Abels og Ruffinis bevis for umuligheten av å løse den generelle *n*'tegradsligningen algebraisk når ≥ 5. *Normat*, 38 (2), 53–84.

Smith, David Eugene (1959) *A Source Book in Mathematics*. Dover, New York.

Sørensen, Henrik Kragh (2004) *The Mathematics of Niels Henrik Abel*. Ph.D. Dissertation, History of Science Department, University of Aarhus.

Stedall, Jacqueline (2011) *From Cardano's Great Art to Lagrange's Reflections*. European Mathematical Society, Zürich.

Stieltjes, Thomas Joannes (1890) Sur la function exponentielle. *Comptes rendus hebdomadaires des séances de l'Académie des sciences* 110, 267–70.

Stillwell, John (1996) *Sources of Hyperbolic Geometry*. American and London Mathematical Societies.

Stillwell, John (2010) *Roads to Infinity*. CRC Press, Boca Raton, FL.

Stillwell, John (2018) *Yearning for the Impossible: The Surprising Truths of Mathematics*. CRC Press, Boca Raton, FL.

Stubhaug, Arild (2013) *Niels Henrik Abel and his Times—Called Too Soon by Flames Afar*. Springer, Berlin.

Toretti, Roberto (1984) *Philosophy of Geometry from Riemann to Poincaré*. Reidel, Dordrecht.

Traub, Joseph (2007) *The unknown and the unknowable: A talk With Joseph Traub*. Edited by John Brockman. http://www.edge.org/3rd_culture/traub/traub_p2.html. Accessed November 5, 2007.

Tropfke, Johannes (1937) *Geschichte der Elementar-Mathematik. Dritter Band: Proportionen, Gleichungen*. Walter de Gruyter, Berlin.

Unguru, Sabetai (1975–76) On the need to rewrite the history of Greek mathematics. *Archive for History of Exact Sciences*, 15, 67–114.

Van der Waerden, Bartel L. (1961) *Science Awakening*. Oxford University Press, Oxford.

Vickrey, William (1960) Utility, strategy, and social decision rules. *Quarterly Journal of Economics*, 74, 507–35.

Viète, François (1593) Variorum de rebus mathematicis responsorum, liber VIII.

Voelke, Jean-Daniel (2005) *Renaissance de la géométrie non euclidienne entre 1860 et 1900*. Peter Lang, Bern.

Volkert, Klaus (2013) *Das Undenkbare denken. Die Rezeption der nichteuclidischen Geometrie im deutschsprachigen Raum (1860–1900)*. Springer, Berlin.

Von Neumann, John (1932) *Mathematische Grundlagen der Quantenmechanik*. Springer, Berlin. English translation 1955.

Wallis, John (1656) *Arithmetica Infinitorum*. Robinson, Oxford. Translations and page numbers are from the English translation by Jacqueline A. Stedall: The Arithmetic of Infinitesimals. Springer, New York, 2004.

Wallis, John (1668) Letter to Lord Brouncker Dated Oxford November 4, 1668. In Huygens (1895), 282–9.

Wantzel, Pierre Laurent (1837) Recherches sur le moyens de reconnâitre si un problème de géométrie peut se résoudre avec la règle et le compas. *Journal de Mathématiques pures et appliquées*, 2, 366–72.

Wantzel, Pierre Laurent (1845) Démonstration de l'impossibilité de résoudre toutes les équations algébriques avec des radicaux. *Nouvelles Annales de Mathématiques*, 4, 57–65.

Waring, Edward (1770) Meditationes Algebraicae. Cantabrigiae. Page references to: *Meditationes Algebraicae: An English Translation*. Edited and translated by D. Weeks. Amer. Math. Soc., Providence, RI.

Webb, Judson (1995) Tracking Contradictions in Geometry: The Idea of a Model from Kant to Hilbert. In Jaakko Hintikka (Ed.), *From Dedekind to Gödel*, 1–20. Kluwer, Dordrecht.

Weil, André (1984) *Number Theory: An Approach through History. From Hamurapi to Legendre*. Birkhäuser, Boston.

Wigner, Eugene (1960) The unreasonable effectiveness of mathematics in the natural sciences. *Communications in Pure and Applied Mathematics*, 13, 1–14.

Winicki-Landman, Greisy (2007) Making Possible the Discussion on Impossibility in Mathematics. In Paolo Boero (Ed.), *Theorems in School*, 183–95. Sense Publishers, Rotterdam.

Yandell, Ben (2003) *The Honors Class: Hilbert's Problems and Their Solvers*. A K Peters/CRC Press, New York.

Zaadz (2007) *Quotes about impossibility*. http://quotes.zaadz.com/quotes/topics/impossibility?page=1-19. Accessed 5/11/2007.

Zach, Richard (2019) Hilbert's Program. The Stanford Encyclopedia of Philosophy, Fall 2019 Edition. Edited by Edward N. Zalta. https://plato.stanford.edu/archives/fall2019/entries/hilbert-program/.

Zeuthen, Hieronymus Georg (1896) *Geschichte der Mathematik im Altertum und Mittelalter*. Høst & Søn, Copenhagen.

Index

Abel, Niels Henrik (1802–1829) 149–53, 170–1
absolute measure of length 191
Ackermann, Wilhelm (1896–1962) 216
actual infinity 205, 216
Al-Asṭurlābī (12th century) 61
Al-Kāshī, Jamshīd (1380–1429) 60
Al-Khwārizmi, Muḥammad ibn Mūsā (c. 780–c. 850) 63, 124
Al-Kūhī (second half of the 10th century) 62, 63
Almagest 49
Al-Samaw'al (c. 1130–c. 1180) 61–3
Al-Sijzī (10th century) 60
alternate currents 130
amateurs 6–9
Ammonius (c. 440–c. 520 AD) 41, 47
Apollonius of Perga (c. 250–c. 190 BC) 39, 44
application of areas 53–5
Arabic mathematics 59–64
Archimedes (c. 287–212 BC) 32–3, 34–5, 40, 49, 59, 61, 63
Archimedes' spiral 33–5, 45, 84
Archytas (c. 380 BC) 37
Argand, [Jean-Robert (1768–1822)?] 130
Aristophanes (c. 400 BC) 28
Aristotelian cosmology 183
Aristotle (384–322 BC) 20, 32, 43, 84, 182, 183, 205, 210–11
Arithmetica Infinitorum 83, 85–9, 99, 121
Arnauld, Antoine (1612–1694) 66
Arrow, Kenneth J. (1921–2017) 251–6
Arrow's impossibility theorem 248, 251–6
Ars Magna 124–5, 127–8, 140
axiomatic deductive method 181–2
axiomatization 209–12, 245
axiom of choice 208–9

Babylon 15–18
Banach, Stefan (1892–1945) 208
Banach-Tarski paradox 208
Barrow, John D. (1952–2020) 9, 10, 247
Bell, John (1928–1990) 246
Beltrami, Eugenio (1835–1900) 194–7, 261
Bentham, Jeremy (1748–1832) 251
Bergson, Abraham (1914–2003) 251, 256
Berkeley, George (1685–1753) 131
Bernays, Paul (1888–1977) 216, 220
Bézout, Etienne (1730–1783) 140, 141
Black, Duncan (1908–1991) 250–1, 253, 263
Bohm, David (1917–1992) 245–6
Bohr, Niels (1885–1962) 243–5
Boltzmann, Ludwig (1844–1906) 244
Bolyai, Farkas (1775–1856) 202
Bolyai, Janos (1802–1860) 191–3
Bombelli, Rafael (1526–1572) 128–9
Borchardt, Carl Wilhelm (1817–1880) 164
Borda count 257
Born, Max (1882–1970) 244
Brahmagupta (c. 598–c. 668) 123
Brāhmasphuṭasiddhānta 123
Brouncker, Lord William (1620–1684) 98
Brouwer, Luitzen E. J. (1881–1966) 215–16
Bryson (fl. late 5th century BC) 43
Buffon, Georges-Louis Leclerc, Comte de (1707–1788) 108

Cantor, Georg (1845–1918) 165–6, 206, 215
Cardano, Gerolamo (1501–1576) 124–8
Cardano's formula 128
Carnot, Sadi (1796–1832) 241
Cauchy, Augustin Louis (1789–1857) 130, 148–9, 233
Chebychev, Pafnuti Lvovich (1821–1894) 179
Chuquet, Nicolas (c. 14.50–c. 1490) 123
Church, Alonzo (1903–1995) 222

classical problems, *see* duplication of the cube; quadrature of the circle; trisecting the angle
classification of impossibilities 12–13, 260–3
Clausius, Rudolf (1822–1888) 241
Clavius, Christopher (1538–1612) 81, 84
Clifford, William Kingdon (1845–1879) 193, 197
Cohen, Paul (1934–2007) 208
Colding, Ludvig August (1815–1888) 241
complex functions 130
complex numbers 121–30
conchoid of Nicomedes 39, 41
Condorcet, Marie Jean Antoine Nicolas de Caritat, marquis de (1743–1794) 75, 76, 79–80, 113, 116–17, 169, 248–9
Condorcet paradox 249
Condorcet winner 249
conic section 37–8, 44–6
continuum hypothesis 207–9
creativity 13–14, 121–30, 263–4
Crelle, August Leopold (1780–1855) 150
cubic equation 60, 63–4, 70, 127–9, 163

D'Alembert, Jean-Baptiste le Rond (1717–1783) 79, 113, 114–17, 130
Davis, Martin (b. 1928) 222
Dedekind, Richard (1831–1916) 166, 237
defect of triangle 190–1, 194
definite quadrature of the circle 90
Degen, Carl Ferdinand (1766–1825) 149
Dehn, Max (1878–1952) 93, 203–4
Delambre, Jean-Baptiste Joseph (1749–1822) 144–8
Delian problem, *see* duplication of the cube
Descartes, René (1596–1650) 65–74, 77, 79–80, 84–5, 107, 129, 141, 245
Determinism 243–5
Dinostratus (c. 390–c.320 BC) 34
Diophantine equations 221–3
Diophantus of Alexandria (c.250 AD) 221, 225, 227
Diorism 52–7, 63
Dirichlet, Peter Gustav Lejeune (1805–1859) 232
Disquisitiones Arithmeticae 145, 156–8
Du Bois-Reymond, Emil (1818–1896) 200
Duhem, Pierre (1861–1916) 239
Dummett, Michael (1925–2011) 257
duplication of the cube 35–9, 44–7, 65–79, 160–3
Du Sautoy, Marcus (b. 1965) 10

École Polytechnique 132, 161
Egypt 15–16
Ehler, Carl Gottlieb (1685–1753) 134
Einstein, Albert (1879–1965) 197, 244
elementary functions 173–4
elementary symmetric polynomials 141–2
elliptic integral 149, 169, 171, 176
energy 238–48
enlightenment 77, 114, 118
entropy 241
equidecomposable 202–4
Eratosthenes of Alexandria (276–194 BC) 35–8
Euclid of Alexandria (fl. 300 BC) 23–6, 29–31, 38, 42–3, 52–8, 61, 182–4, 202–3, 205, 210, 225
Euclid's algorithm 21–3, 119
Eudoxos (c. 408–c. 355 BC) 24, 31, 37
Euler, Leonhard (1707–1783) 93, 117, 119, 129, 130, 131, 132, 134–9, 140–1, 158, 168, 229
Euler tour 135–6
Eutocius (c. 480–540 AD) 38, 44, 47, 49
existence and constructability 42–4

Farquharson, Robin (1930–1973) 257–8
Fermat, Pierre de (1601?-1665) 66, 158, 212, 225–30
Fermat's last theorem 225–37
Fine, Oronce (1494–1555) 81
Fontaine, Alexis (1704–1771) 169
foundational crisis 215–16
Fraenkel, Abraham (1891–1965) 208, 215
Fremat prime 158
French Academy of Sciences 75, 81, 108, 116, 131, 173, 232–3
Frénicle de Bessy, Bernard (c. 1604–1674) 230
Frey, Gerhard (b. 1944) 235–6
Fuchs, Lazarus (1833–1902) 178

Galilei, Galileo (1638, 31–33) 206
Galois, Evariste (1811–1832) 153–4
Gauss-Bonnet theorem 194

Gauss, Carl Friedrich (1777–1855) 130, 145, 155–60, 190–4, 202, 205, 230–1
Gaussian curvature 194
Gelfond, Aleksander (1906–1968) 205
geometric algebra 56, 126
Germain, Sophie (1776–1831) 230–2
Gerwien, Paul (c.1800–1858?) 202
Gibbard, Allan (b. 1942) 256–8
Gibbard-Satterthwaite theorem 256–8
Girard, Albert (1595–1632) 129, 141
Gödel, Kurt (1906–1978) 217–21
Gödel's incompleteness theorems 217–21
graph theory 133–9
Greek mathematics 19–59
Gregoire de Saint-Vincent (1584–1667) 81–2
Gregory, James (1638–1675) 89, 91–9, 101, 106

Hardy, Godfrey Harold (1877–1947) 179
Heisenberg, Werner (1901–1976) 243–4
Helmholtz, Hermann von (1821–1894) 193–4, 197, 241
Hermann, Grete (1901–1984) 245
Hermite, Charles (1822–1901) 164–5
Hertz, Heinrich (1857–1894) 245
Heuraet, Hendrick van (1634–1660?) 84
hidden variables 244–7
Hierholzer, Carl (1840–1871) 138–9
Hilbert, David (1862–1943) 153, 166, 195, 200–5, 207, 209, 210–16, 220–4, 236, 238
Hilbert's first problem 205–9
Hilbert's Paris problems 199–209, 212–15, 221–3
Hilbert's second problem 212–15
Hilbert's seventh problem 204–5
Hilbert's tenth problem 221–3
Hilbert's third problem 202–4
Hippasus of Metapontum (5th century BC) 19–20
Hippias of Elis (fl. 420 BC) 34
Hippocrates of Chios (c. 470–c. 410 BC) 30–2, 36–7, 44
Hobbes, Thomas (1588–1679) 81
horocycle 192
horosphere 192
Houël, Guillaume-Jules (1823–1886) 195, 198
Hurwitz, Adolf (1859–1919) 166
Huygens, Christiaan (1629–1695) 83, 95–9, 105, 106

Iamblichus (c. 245–c. 325 AD) 19
Ibn al-Haytham (c. 965–c. 1040) 60–1, 185–6
ideal elements 121, 130–2
ignorabimus 199–201, 223, 238
impossibility in natural science 10–11, 238–47, 262
impossibility theorem, meaning 4
incommensurability 19–26
incompleteness 216–21, 223–4
indefinite quadrature of the circle 90
independence of axiom 213, 261
indeterminism 243–4
indirect proof, *see* proof by contradiction
infinitesimal 130–2
integral 148–9, 168–79
integration in finite terms 168–79
intuitionism 215–16
irreducible case of cubic equation 72, 128–9, 162–3
Islamic mathematics, *see* Arabic mathematics

Joule, James Prescott (1818–1889) 241

Kant, Immanuel (1721–1804) 193, 197
Kelvin, Lord, *see* Thomson, William (Lord Kelvin) (1824–1907)
Klein, Felix (1849–1925) 159, 162, 165–6, 178
Knuth, Donald (b. 1938) 258
Königsberg bridges 133–9
Königsberger, Leo (1837–1921) 151, 178
Koopmans, Tjalling Charles (1910–1985) 255
Kronecker, Leopold (1823–1891) 207, 215, 237
Kummer, Ernst (1810–1893) 233, 236

Lacroix, Sylvestre François (1765–1843) 146, 148
La Géométrie 66, 77
Lagny, Thomas Fantet De (1660–1734) 110–12

Lagrange, Joseph-Louis (1736–1813) 132, 140–7, 222
Lambert, Johann Heinrich (1728–1777) 118–20, 163–4, 177, 189–90
Lamé, Gabriel (1795–1870) 232–3
Langlands, Robert (b. 1936) 235
Laplace, Pierre-Simon Marquis de (1749–1827) 170, 243–4
Legendre, Adrien Marie Legendre (1752–1833) 156, 160, 170, 176, 190, 230–2
Leibniz, Gottfried Wilhelm (1646–1716) 83, 85, 99–100, 105–6, 129, 131–2, 134, 139, 168, 183, 240
Leibniz's infinite series for π 83
Leonardo da Vinci (1452–1519) 240
Leonardo Fibonacci da Pisa (c. 1170–c. 1250) 123
l'Hospital, Guillaume François Antoine, marquis de (1661–1704) 109
Liber Abaci 123
Lindemann, Ferdinand (1852–1939) 164–7, 204
Liouville, Joseph (1809–1882) 91, 163–4, 169, 171–8, 233
Liouville's principle 174
Liouville's theorem 174–5
Lobachevsky, Nikolai Ivanovich (1792–1856) 190–4, 261
Longomontanus, Christian (1562–1647) 81
Ludolph van Ceulen (1540–1610) 81, 83

manipulation of voting procedure 256–8
Marinus (late 5th century) 47
Matiyasevich, Yuri (b. 1947) 222
Maxwell, James Clerk (1831–1879) 240, 241, 244
Mayer, Robert von (1814–1878) 241
mean proportional 46
Menaechmus (c. 350 BC) 37–8
Mersenne, Marin (1588–1648) 65, 81, 84
Mesopotamia, *see* Babylon
meta mathematics 5–6, 8–10, 17, 58–9, 65, 76, 144, 163, 180, 260–1
Minkowski, Hermann (1864–1909) 199
Montucla, Jean Étienne (1725–1799) 76–80, 113–14, 117–18, 143

Neile, William (1637–1670) 84
neusis construction 40–1, 45, 59, 61, 263
Newton, Isaac (1643–1727) 76, 99, 102–5, 106–7, 114, 116, 131, 142, 240
Nicholas of Cusa (1401–1464) 81
Nicomedes (c. 280–c. 210 BC) 34–5, 39, 41
Nobel Memorial Prize 255
Noether, Emmy (1882–1935) 241–2
non-Euclidean geometry 181, 187–98, 212, 261, 263
non-standard analysis 132
number concept 85, 121–4

Oenopides (mid-5th century BC) 30
Oldenburg, Henry (c.1619–1677) 96
Omar Khayyam (1048–1131) 63, 187
Ostrogradsky, Mikhail (1801–1862) 179

Panckoucke, Joseph (1736–1798) 108
Pappus of Alexandria (c. 290–c. 350 AD) 19, 44–7, 50, 56, 58–9, 67, 168, 260
parallel postulate 181–90, 192–3, 195, 198, 201, 212, 239, 261
Pell John (1611–85) 81
pentagram 21, 22
perpetual motion 75, 108, 116, 238, 240–1, 262
Petersen, Julius (1839–1910) 159, 161–2
Philoponus (fl. c. 550 AD) 43, 61
Physics 9, 130, 197, 238–47, 262
π 28, 33–4, 49, 60, 81, 83, 85–9, 90, 99–100, 102, 111, 118–20, 163–7, 201, 204–5
Picard, Émile (1856–1941) 178
plane problem 44–5, 67
Plato (c. 425–c. 348 BC) 35–6, 38, 58, 214
Plimpton 322 18
Poincaré, Henri (1854–1912) 198, 199, 239
Posidonius (c. 100 BC) 185
potential infinity 205
Principia 102, 103, 216, 217, 219
proof by contradiction 20
Ptolemy (c. 100–c. 170 AD) 48, 49–51
Putnam, Hillary (1926–2016) 222
Pythagoras (c. 570–c. 495 BC) 19, 20
Pythagoras' theorem 4, 9, 20, 23, 31, 50, 184, 225
Pythagoreans 21, 28, 37
Pythagorean triplet 17, 225, 227–8

quadratic equation 17, 53–5, 63, 66–8, 74, 124, 127, 155, 158–9
quadratrix 34–5, 39–40, 45, 47, 81, 84, 109
quadrature of the circle 28–35, 47–8, 81–120, 163–7, 201
quantum mechanics 243–7
Quine, Willard Van Orman (1908–2000) 239

RAND corporation 253
reciprocal table 16
rectification 32, 34–5, 43, 47–9, 84, 101, 109–12
regular polygon 61–2, 155–61
relativity theory 197, 242, 247, 256
Ribet, Kenneth (b. 1947) 235
Riemann, Bernhard (1826–1866) 130, 193, 196–7
Risch, Robert Henry (b. 1939) 177, 179
Ritt, Joseph Fels (1893–1951) 177, 179
Robinson, Abraham (1918–1974) 132
Robinson, Julia (1919–1985) 222
Rosenlicht, Maxwell Alexander (1924–1999) 179
Ruffini, Paolo (1765–1822) 145–9, 150–3, 263
ruler and compass 30, 155–61
Russell, Bertrand (1872–1970) 211–12, 215–16, 220
Russell's paradox 215

Sacccheri, Girolamo (1667–1733) 187–9
Samuelson, Paul A. (1915–2009) 251, 256
Saurin, Joseph (1659–1737) 108–10
Scaliger Joseph Justus (1540–1609) 81
Schneider, Theodor (1911–1988) 205
Schwarz, Hermann Amandus (1843–1921) 166
Scipione del Ferro (1465–1526) 127
Serre, Jean-Pierre (b. 1926) 235
set theory 205–9, 215–17
sexagesimal system 16–17, 60
Sharaf al-Dīn al-Ṭūsī (1135–c. 1213) 63–4
Shimura, Goro (1930–2019) 235–6
Simplicius (about 490–560 AD) 47, 92
Sluse, René-François de (1622–1685) 76
social choice 248, 251–8
solid problem 44–5
solution by radicals 88–9, 92, 101, 140–53, 157, 163, 168, 171–2, 179, 201, 259, 261
square root of 2 20–1, 23
squaring the circle, *see* quadrature of the circle
Stieltjes, Thomas Joannes (1856–1894) 166
strategy-proof voting procedure 256–8
Sydler, Jean-Pierre (1921–1988) 204
Sylvester, James Joseph (1814–1897) 137–8

Taniyama-Shimura conjecture 235–6
Taniyama, Yutaka (1927–1958) 235
Tarski, Alfred (1901–1983) 208, 252
Tartaglia, Niccolò (1500–1557) 127
Taylor, Richard (b. 1962) 236
Thābit ibn Qurra (2. half of the 9th century) 59
Thales (c. 600 BC) 19, 29, 31
Theon of Smyrna (fl. 100 AD) 35
Theorema Egregium 194
Thomson, William (Lord Kelvin) (1824–1907) 241
time travel 11, 247
transcendental number 93, 94, 102, 119–20, 164–7, 204–5
transitive ordering 249, 253
Traub, Joseph F. (1932–2015) 10
trigonometry 50–1, 59–60, 192
trisecting the angle 34, 39–41, 43, 45–7, 50–1, 56, 59–61, 69–80, 84, 98, 101, 108–9, 112, 155, 161–3, 167, 180, 220
Turing, Alan (1912–1954) 219, 221, 222
two mean proportionals 36–9, 43, 45–9, 70–4, 75–6, 78, 80, 92, 112

Ulūgh Beg (1394–1449) 60

Vandermonde, Alexandre-Théophile (1735–1796) 140
Vessiot, Ernest (1865–1952) 179
Vickrey, William S. (1914–1996) 258
Viète François (1540–1603) 66, 72, 83, 90, 102
Von Neumann, John (1903–1957) 215, 216, 219, 245–7
voting 248–51, 252–8, 263

Wallace, William (1768–1843) 202
Wallis, John (1616–1703) 81, 83, 85–9, 90, 96–9, 121–3, 127, 186, 230, 259, 263
Wallis' product for π 83, 90

Wantzel, Pierre Laurent (1814–1848) 151–2,156, 158–9, 160–3, 180, 202, 259
Waring, Edward (1734–1789) 140, 141
Weierstrass, Karl (1815–1897) 130, 166
Weil, André (1906–1998) 228, 235
welfare economics 248, 251–6, 263
well-ordering theorem 207–9
Wessel, Caspar (1745–1818) 130
Wigner, Eugene (1902–1995) 130
Wiles, Andrew (b. 1953) 234–6
Wolfskehl prize 234, 236
Wren, Christopher (1632–1723) 85

Zermelo, Ernst (1871–1953) 208, 215, 217